Are You Ready to Change the World?

Thoughts on Technology Leadership for the Future

By S A James

DEDICATION

To the next generation, lead with heart and lead with passion.
Elizabeth and Matilda lead your way.

Always remember:

"One voice, can change a room, and if one voice can change a room, then it
can change a city, and if it can change a city, it can change a state, and if it
can change a state, it can change a nation, and if it can change a nation, it
can change the world. Your voice can change the world."

Politician and Attorney - Barak Obama

CONTENTS

ACKNOWLEDGMENTS

The world is a beautiful place when people pause and take time for others. To all those who stopped their normal routine to dedicate time to this project, thank you. To those who took time to mentor me throughout my career, thank you. To those who strive to help others learn to grow, thank you. Leading the leaders is a legacy anyone would be proud of, and to those who take up the challenge, I thank you. To all the individuals I've had the opportunity to lead, be led by, or watch their leadership from afar, I want to say thank you for being the inspiration and foundation for this book.

I must start by thanking my awesome husband, Robert. From hearing early drafts late at night to giving me advice on the cover to keeping the kids out of my hair so I could edit, formulate, rewrite, draw and think, he was as important to this book getting done as I was. Thank you so much, beautiful man.

Thank you to my amazing family, Mum, Dad, Katie, Ed and Christine. Thank you for your never-ending support, ideas, encouragement, guidance, wisdom and above all believing in me, Robert and our crazy ideas. Someone has to have them; why not us?

Thank you to the friends that became family, the Elliots and the Martins and those on the lane. You helped us to be OK with thinking differently, with living a different kind of life, and for that I thank you greatly.

To my favourite mentors over the years, Mal, Rollo and Rob, Jacqui, Marek, Mark H and Mark B, Paul, Mandy, Frank, Phil, Sharpie, Courtney, Heather, Penny, Jacques, Anna, Tony, Lance and more recently Des, you helped me to grow and continue to do so. To John, the submarine commander, and Kitty, both who are no longer with us -- you both taught me so much when I was so young. Thank you.

Without the experiences and support from my peers and team at DXC and the CSC legacy group, this book would not exist. Thank you specifically to Des, Kevin and Seelan for giving me another chance and letting me fly, to come back when I was ready. I'm eternally grateful and proud to see that loyalty runs deep. You have given me the opportunity to lead with a great group of individuals. To be a leader of great leaders is a blessed place to be. Thank you to my team for asking the questions, wanting to grow and for pushing the boundaries of what we thought possible. To my fellow leadership team, thank you for being there.

Thank you also to my Facebook, Twitter and Linkedin connections. Without you, my voice would have stayed silent. Thank you to the Women in Technology Western Australia (WiTWA), women who look after each other and care. Thank you Annu, Ro, Lacey, Donna and Annie, virtual friends that have inspired me to go beyond and have a go. Thank you to The Autism Academy for Software Quality Assurance (AASQA), particularly to the board that accepted me for me and welcomed, once again, my crazy ideas with encouragement and sincerity. To the college students and graduates that I have mentored over the years, remember you taught me as much as I taught you.

Thank you to all of those who wrote books that inspired me and helped me see different ways to lead. Thank you for having the courage to write, as to do so is to pour your soul into the book. Thank you to all those who write music that inspires me. "If music be the food of love play on," said Shakespeare, and I agree. Every word written in this book was written to music.

Thank you to Josie for taking my ideas for a book, understanding me and making the cover pop. Thank you to Christine, my editor, who has helped to keep me honest, tidied up my messy scrawl and stood by me for over 6 years. Christine, this book would not be possible without you and I thank you from the bottom of my heart for editing my raw notes and for probing for more detail when times got hard and writing got tough.

Finally, thank you to Troy, our dog who listened to everything without a word or a woof, and especially to Elizabeth and Matilda. Without you, I would never have wanted to write this book. You are the inspiration for Mummy to change the world, one word at a time. This book turned out to be more than I could ever have imagined, and I hope it inspires you, too.

Sarah James

Preface

This book is about changing the world, being a rebel and having a go. It's about how we recognise the need for shifts and summon the courage to make those changes, starting at the smallest level and eventually revolutionizing the very way we lead our teams, our companies and our world. I've often had people, in work and in life, tell me it seemed I was trying to change the world. I would get ahead of them and, sometimes, myself with wild ideas, and they'd encourage me to slow down, break it down so they could understand where my brain was headed. This would frustrate me when I was younger, but not so much now. Now I'm proud to think big -- and changing the world seems a great place to start. Hence the title for this book.

As one of Australia's leading female technologists, I've spent my career looking to the future. I've helped more than 40 companies adjust to the information revolution of the last 20 years. I've supported business leaders looking for ways to survive the turmoil of the digital age and position their companies for growth and prosperity, with an ever-watchful eye toward what's next. Along the way, I've learned a little something about leadership, about what society needs to achieve to be, not only successful, but sustainable. It's clear to me now, as reports about climate change, geopolitical turmoil and global pandemic fill our daily news feeds, that the future is no longer a given. Change at all levels must happen to ensure we even have a tomorrow, let alone the bright one we have the potential to create if we combine good leadership with the amazing technologies in development right now.

It's my hope that the insights I provide here, based on my personal experiences and many years working in the field, will help prepare and inspire you to face the challenges the present demands. Now more than ever, we need to lead radically, think differently and bring together diverse teams of people with a range of personalities and perspectives to make real change. Let this book guide you in thinking about the world today and the world we could create tomorrow -- one unlike anything we've known before, where humans and technology become even more closely entwined and a new type of leader answers the call.

Chapter 1

Are You Ready to Change the World?

"I'm so sick of running as fast as I can.
Wondering if I would get there quicker if I was a man."
"The Man" by Taylor Swift

A colleague, a fellow female leader who I find to be extremely courageous and much more of a rebel than I could ever be, recently shared the Taylor Swift song "The Man" with me. She knew I needed a little inspiration to get through a difficult day, and Taylor Swift seems to always have the right song for that. In "The Man" Ms. Swift sings about how easy success seems to come to the men around her. She lays bare how society doesn't question powerful men in the same way it does powerful women. "They'd say I hustled, put in the work," she sings, about how the world would react to her success if she were a man. "They wouldn't shake their heads and question how much of this I deserve." My

1

colleague recognised that I would love the song, and she was right. Ms. Swift, an incredibly powerful young woman and leader in her own right, put to music a lot of the same ideas I've been pondering in recent years.

If I were really honest, those questions -- and, in fact, this book -- started swirling in my head years ago when I was writing a blog post in advance of the 2016 U.S. election. The post I had written was pondering the possibilities of the first female president of the USA, a call for young girls everywhere to be inspired by another glass ceiling shattered. My editor, Christine, loved the piece, and we had it ready to publish the morning after the election, depending on the results, of course. The next day our hearts, like those of many women around the world, sank over Hillary Clinton's loss and the thought of what a Donald Trump presidency might mean for the world. The hopeful words of my blog suddenly became meaningless, and we felt that publishing the piece on our company blog would be controversial. However, In an alternate universe -- one in which I had more fearlessness myself -- I would have published the blog post anyway. Though I can't go back in time, Sarah 2.0 *can* share those words I wrote in 2016, with a few updates here and there, in this space. I believe its time has come:

The Hidden Chapter

We Wait and Wait with Bated Breath

Written 2016

As you're sure to be aware, the U.S. may elect Hillary
Clinton as the country's first female president tonight. I've been
lucky enough in my life (whist living in the UK and Australia) to
have been led by two female leaders, Margaret Thatcher and Julia
Gillard. (I also watched from afar as Theresa May led the UK
during the particularly turbulent time period of 2016-2019.) These
women didn't always find success in their political causes, but they
definitely made an impression -- and they continue to inspire
female leaders today. Ms. Gillard, still very much a prominent
leader in Australian society, fought for a carbon tax scheme that
was radical at the time. And when it comes to politics, timing is
everything. Ms. Thatcher was incredibly disruptive serving as Prime
Minister in an era when it was not popular for a woman to be in
charge. When she passed away years later, people remembered her
as "good" or "bad" depending on which side of the political fence
they stood during her rule. What these female leaders (and Ms.
May, as well) often endured was, in my opinion, terribly hard and

3

sometimes soul destroying. Alas they kept up the fight for what they believed in, reminding me of the famous words of poet Dylan Thomas, "Do not go quietly into that good night. Rage, rage against the dying of the light."

One of my dear friends, a mentor and male engineer, questioned me about the significance of Ms. Gillard's 2010 election in Australia: "Was her election seen as a turning point in women's rights in your country?" he asked. "Did you see Ms Gillard's reign as groundbreaking, something you can show your daughters and say, 'See you can be anything you want.' Is Ms. Gillard a role model for people?" This was a fascinating question, as rarely did I think about the women who served as role models in my life outside of my family. I always look to my mother, my grandmother, my sister, my sisters-in-law, some very dear family friends and, more recently, my mother-in-law as women who taught me over the years. But I've undoubtedly also been influenced by people I've never met but knew through the media, people like Madonna, Whitney Houston, The Spice Girls, Princess Diana -- and, yes, Margaret Thatcher and Julia Guillard.

My sister and I grew up during the years of the Iron Lady, and our family raised us with the mentality that we could be anything we wanted to be. We grew up thinking that success was based on the merit of who we were as people, not on any labels given to us and certainly not on our gender. This way of thinking led to both of us being leaders who judge people in the workplace based on what they produce, not by the colour of their skin or if they're a woman or man. As a family, we were encouraged to discuss and understand the effects of Ms. Thatcher's ideas and sometimes controversial policies on our society. We were encouraged to discuss both sides, our viewpoints and also our biases. As members of Young Farmers, a UK organisation that teaches rural kids about agricultural practices, public speaking and leadership, we saw the effect of certain policies, such as mine closures, on our local villages or the reform of the common agriculture policy, the effect of foot and mouth on our fellow farmers and the effect of BSE on the economy. We saw how Thatcher's ideas affected people first hand. My family and friends of ours became split on different sides of the picket lines over different issues. Politics and leadership became real and exciting, if controversial, under the formidable Mrs. Thatcher.

With this background, I began to think about Ms. Gillard's 2010 election in Australia, wondering if it made as much of a difference in my life. I think it did. I remember standing with one of my dear friends and colleagues in a client office, watching in awe as Ms. Gillard was elected. We both knew it would be an interesting time, especially given the way she was elected, and we stood in silence, excited by what might come. I found Ms. Gillard to be a great leader -- and I think she still is. But she endured some really terrible treatment during her years in power. She was often spoken over, her outfits critiqued more than her policies and her relationships with friends questioned to the point of people wondering about her sexuality. Still, in the face of criticism and controversy, she held her head high when others were rude and untrustworthy, and she represented Australia well. Her election was one of the best things that could have happened to our country though it wasn't seen as a turning point for women's rights at the time. Looking back now, I can see that it did seem to encourage women in leadership roles, including myself. Will Ms. Clinton do the same? Watching from afar, I've seen Ms. Clinton show amazing strength, poise, grace and calm when dealing with the opposition. And I believe she has now paved the way for many, many more women to throw their hats in the ring. (A note from my future self: The mid-term elections of 2018, resulting in historic wins for American women, certainly proved that to be true.) Personally, I'd love to see Michelle Obama put her hat in the ring.

While there are many cultural issues in the U.S. that distinguish the country from my experiences in the UK and Australia, I am certain any country would be hurt by a leader like Donald Trump, someone who disregards people of different religions, women, treaties, even honesty. Leaders succeed when they put other people at the forefront. Decisions are best made with the mindfulness of Sherlock Holmes and full consideration of others. The guiding question of any President's decision should be: "How will this affect our people?" A leader wants to be judged on how he or she leads and the good he or she does. And a good leader, in turn, should judge others in the same way, treating them with integrity, treating others the way he or she wishes to be treated. As Americans decide who will lead their country in the years ahead, I watch and wait with anticipation from the other side of the world, hopeful that another strong female leader will step onto the stage and, in Ms. Clinton's words, "harness the power of technology and innovation so that it works for all Americans."

Are You Ready to Change the World?

4 years later...

Much has changed since I wrote those hopeful words four years ago. There has undoubtedly been a wave of female empowerment, with CEOs committing to gender equality, the #MeToo movement changing entire industries and the fight for pay equity generating headlines around the world. But the American election, in some ways, felt like a giant step backward in terms of female leadership and, in fact, the very style of leadership celebrated in politics and business. Leading with compassion, honesty and heart seemed to be pushed aside by people who favor dishonesty, selfish motives and, even, bullying. Well I believe now is the time to push back. Any of us can learn to be a fearless leader, no matter our gender, our experience, our background. We can release our worries about being different, focus on making a difference and just jump in. It's time to break down the barriers in your own mind -- and the minds of others -- and lead with heart, above all.

In the last four years, I've been reading about leadership and compassion, listening to podcasts and TED Talks and absorbing ideas and information from colleagues in Australia and around the world. At the same time, I've watched surprises on the world stage, like New Zealand Prime Minister Jacinda Arden coming to power and showing us all the strength that can come from kindness. I've seen teenager Greta Thunberg, an environmentalist from Sweden, represent her generation in a powerful fight against climate change, putting into words what many of us have been thinking but unable to voice. In the technology space, people like Facebook's Sheryl Sandberg and journalist Rebecca Holman have written female leadership books for alphas and betas alike. All of these ideas have percolated and interacted and come together here. In writing this book, I feel I'm talking to my own two daughters (ages 5 and 6) and others like them, hopefully showing them that in 20 years' time, when they graduate from university, some 40 years after I did, they will see that leading with heart can be scary, but it's possible. And, in fact, it's a necessity in the future that lies ahead of us. Technological developments, such as Artificial Intelligence (AI), Cyborgs and

Human-to-Computer Interfaces (or, as my 5-year-old calls them, robots) will require an entirely new type of leader, one who can lead teams built of people *and* technology. This type of leadership requires strategic thinking and lateral thinking, soft skills, such as optimism, growth mindset and resilience, as well deep insight into the technical elements involved. Anyone can develop these qualities, I truly believe, but I especially want to encourage young women and girls who, like me, can take their passion for technology and turn it into a meaningful career.

Making change in our communities, organisations and society at large is a challenge that takes time and patience. It requires change makers to be brave, bold, and, above all, kind -- and it requires us to learn from the experiences and mistakes of our past. Fearlessness is one value that can't be coded into a human, though I sometimes wish it could. We need to not be afraid to speak up when we see something that isn't right. We need to be courageous and have those dreaded, difficult conversations that clear the way for something better and more promising to come. Eventually, we will be having these conversations with robots and AI, not just our human colleagues in the workplace. Imagine having to switch off a robot that has become redundant, replaced by a better version of itself. Imagine if that robot was a blend of human characteristics and technology. In cases like this, we will need to have rebellious conversations with ourselves, to ensure we are doing the right thing. In fact, our moral code as human beings may need to be entirely rewritten to accommodate the technological advancements of the time.

If I were entering the technology space today, I would tell my younger self "Do not be afraid of a title; everyone is somebody." I have been talking to people in the C-Suite, since my early 20s with a humble naivety that has allowed me to be boldly honest with some leaders, a necessity to become a trusted advisor. I would also say, "Reach out to those females around you. If you can see her, you can be her." Last year I had my first all-female meeting and my first all-female team -- after 20 years of working! Sometimes we can go too far in the other direction, we need to be considerate of this, but in all aspects of work, we should ensure diversity in age, gender, viewpoints, etc. In order to counter our biases, we need to ensure that we have multiple viewpoints at the table and that they are heard. I believe the leaders of the future will break rules, speak up, lead with heart, be fearless and be open to change. I believe they will "Start small, dream big and get stuff done." And I believe they will be you, me and all of us getting ready for the world to change -- and eager to help make it happen.

When I think about the future and where we are going, it all comes down to this simple sentence: "The best work of your life is the most meaningful." I believe this work will be our most meaningful yet -- and this book will offer some authentic tips from my own experiences to propel you on your own leadership journey. Thank you for reading. You are awesome. Now, are you ready to change the world?

TAKE ACTION

As you read this book, I would like to encourage you to take small actions to get you thinking and acting in a new or different way. You will find these "Take Action" ideas sprinkled throughout the book. When you come across one, consider pulling out a pen and paper or a notes app on your phone and jotting down a few ideas. You could even create a board to track your actions or a map for development.

TAKE ACTION

Courage can be inspired by those we admire. Make a list of people you look up to and consider change-makers in the world. Think about people in your life or throughout history who have kept going when the road gets tough. As Nelson Mandela once said, "Courage is not the absence of fear, but the triumph over it. The brave man is not he who does not feel afraid, but he who conquers that fear."

Chapter 2

Why Diversity Matters, Even to Superheroes

"Sooner or later, we must learn to judge each other on our own merits.
Sooner or later, if man is ever to be worthy of his destiny, we must fill out hearts with tolerance."
-- Marvel Comics Creator Stan Lee

People are fundamentally different. As my mother likes to say, "If everyone were the same, life would indeed be boring." I tend to agree, though I do wonder if my mother borrowed her wisdom from Marilyn Monroe. Ms. Monroe's credited with saying, "Imperfection is beauty. Madness is genius, and it's better to be absolutely ridiculous than absolutely boring." The famous American actress was certainly not boring. A bookworm and extraordinary beauty who suffered from anxiety and depression, even though she was one of the most adored people in the world, she lived a life that was, if nothing else, *different*.

We live in a world and time that likes to pay lip service to the idea of diversity, of being open to differences and accepting of

13

a variety of backgrounds and viewpoints. But knowing the value of diversity and being truly tolerant of others tend to be two very disconnected acts. One of my personal superheroes knew how to talk, the talk *and* walk the walk. He created written worlds where different was celebrated. He wrote all of us -- women, minorities, weirdos and misfits -- into the story, so that we could see ourselves represented in art, first in the comics and then on the big screen. He wrote because he believed diversity mattered, and he believed that everyone should have a favorite superhero to represent them.

Stan Lee, creator of Marvel Comics, inspired generations of fans who loved the worlds and characters he created. He helped us believe that anyone could be a superhero and that we all have a little bit of the extraordinary inside us. Someone once asked me why I chose the Twitter handle @GeoSuperGirl to represent my online persona. The truth is, I love superheroes and heroines, and I wanted to channel that energy in my professional life, which has, at times, focused on geographic information systems. So that's how it came about, and it's since become part of my identity, my alter ego. One thing I love about Stan Lee is the female characters he brought to life through his work. He helped to create women like Jean Grey-Summers, a mutant known as Phoenix with superhuman abilities; Peggy Carter, the WWII codebreaker who helps Captain America fight the Secret Empire; and the more recent She Hulk, a cousin of the original green giant who keeps her emotional cool when she gains super strength. I love that these women have complex backstories and rich character arcs, as well as powers that can be used for the good of others, even if their abilities sometimes cause challenges in their personal lives.

Ironman's Pepper Potts strikes a chord with me because of her innate sense of knowing how and when to do the right thing. She's been forced to get herself and Ironman Tony Stark out of difficult situations more than a few times. Potts is able to be diplomatic when Tony is abrupt. She's humble when Tony brags. She manages the day-to-day routine in a way that allows the superhero a degree of flexibility to be who he is. She's focused and smart and a guiding, calming force for Tony. She knows things others don't and even helps J.A.R.V.I.S, the Artificial Intelligence home computing system, learn how to do some tasks. Although she doesn't appear to have superhuman abilities, I believe Pepper's superpower comes from helping Tony be the best person he can possibly be. *She* is his superpower. And though her behind-the-scenes role may be seen by some as anti-feminist, there are many women in the real world who can relate to the role Pepper plays, who know the honour of being the hero supporting the hero. At home and in the workplace, these individuals guide their families and keep the leaders of their organisation on balance. These are the type of people you definitely want to have represented on your team.

Another character I absolutely love is Captain America's Peggy Carter. She has a way with words and the ability to stand up for herself, a trait many of us would love to call on with her exactness and biting wit. She sometimes lacks a filter, like in the "Captain America: the First Avenger" movie when she tells her recruits, "Faster, ladies! Come on! My grandmother has more life in her, God rest her soul." But it's her fearlessness in crossing barriers and saying what she thinks that makes her such a cool character. Being an agent certainly helps, and, well, she is Captain America's girlfriend. There's this wonderful scene in one of the films in which Agent Jack Thompson mistakes Ms. Carter for a secretary, something I've had happen to me quite a few times over my 20-year career. "If you don't mind, these surveillance reports need to be filled and you're really so much better at that kind of thing," Agent Thompson tells her. To which she sharply replies, "And what kind of thing is that, Agent Thompson? The alphabet? I could

teach you. Let's start with the words beginning with A." Ms.

Carter's character has a fascinating backstory, including a stint at

the famous, real-life Bletchley Park (we were lucky enough to visit a

few years ago, a present from my sister), a once top-secret

headquarters for female code-breakers in Great Britain during

World War II. I sometimes wonder if she was based on a real-life

character, but I guess we will never know. What I do know is that

she is a superhero in her own right, fighting evil with a passion and

confidence most of us would love to have. "I know my value;

anyone else's opinion doesn't matter," she once said. What an

inspiring message for other female leaders, to be brave in the

knowledge of their own worth. I've found this to be difficult to do

at times in my career, but it's a good reminder that knowing your

value can inspire confidence and help you make the best decisions.

Stan Lee did so much to promote diversity in his work. He

didn't shy away from characters that crossed boundaries, and he

spoke openly about the need for human beings to treat each other

with more respect and tolerance. In a 2016 interview about the

Black Panther movie, he said, "If my books and my stories can

change that, can make people realize that everybody should be equal and treated that way, then I think it would be a better world." He had a humour both adults and children could understand, and for some reason, many of his characters resonated with us "geeks and nerds" who went on to careers in technology. In his words, "I was just trying to write stories that people of all ages and sexes would enjoy reading." What he didn't realise was that he was also creating an escape hatch for those of us who did not feel we fit into the mainstream society. He gave us the Marvel Universe as an escape and an inspiration. His world -- accepting of mutants and superheroes, people who use technology to acquire superhuman abilities and people who support the heroes among them -- definitely inspired me to be open to diverse ideas, accepting of the differences in people and aware of the superhero within.

Becoming a parent was one of those moments I felt my superhero rise up. Parents seem to grow powers we did not know were even possible. We find the ability to be patient, the ability to answer many questions at once. We tap into superhuman strength to overcome fatigue and do the dirty work of everyday life in order to make sure our children have their needs met. We gain the ability to relearn what we thought we knew, to love with selflessness and develop a second sense or intuition when something isn't quite right, when the house is just a little too quiet. I was once told that I should never try to be a SuperMum, but I completely disagree with this message. I think parenthood calls on us to go beyond ourselves and reach beyond our boundaries, this also applies to being a leader and leading a team. Until we try, we never know what we are capable of. I experienced this from the very beginning of parenthood. It wasn't meant to be easy for my partner and I to have children, and when we found out we were pregnant with our first, I didn't really know what to do. I had decided to focus on my career thinking parenthood wouldn't be an easy option, so my brain had to rewire quickly to get my head around the prospect of becoming a parent. My better half was over the moon and excited by our happy surprise. But I had to think, could I do it all? Did I have the energy to do it all? Did I know anyone else who did? Two weeks before my first child was born, and one week into my maternity leave, I missed a step walking to the garden and dislocated my leg, breaking it in three places. Over the next few weeks, I was in a wheelchair, taking daily injections and also caring for a newborn baby. Money was tight. We had the help of the Red Cross, which was invaluable, but I only received six weeks of maternity pay, which was paid when I returned to work after 3 months. Thank God for the savings we had put together and my mother ensured we had. We had to sell several new-build rental units we owned to survive financially during my maternity leave, and we took another financial hit with a large tax bill. One day we had less than $50 in the bank. I remember thinking something had to come from somewhere, and suddenly a friend of ours turned up with groceries as a gift. A dear friend said to me once "when you are at rock bottom something will come from nowhere". I went back to work after four months, and the week after I returned to the office, we found out we were pregnant with our second child. I cried tears of sadness, not joy. At the time, I was looking forward to getting my career back on track and getting our family financially stable once again. Eventually those tears turned to smiles as we celebrated the arrival of our second daughter. Through it all, I found a way to tap into my inner superhero and do what had to be done to take care of our growing family.

While becoming a mother seemed to naturally bring out my superhero side, tapping into those powers in my career has been a learning process. One of the ways I've gained an edge was in being open to differences, just as the Marvel Universe taught me to be. I've learned that the key to difference is understanding. If we can understand what we think differently about and *why* then we can more easily find common ground. In technology, differences in backgrounds and biases often add value, something we call market differentiation. We can create better solutions and better products when we have more ideas and better ideas to work with. We can make our solutions more accessible and useful to more customers if we consider the needs of different types of people during design and production. Ask for volunteers, get their feedback and ensure the product is developed with feedback in mind. The earlier you bring these champions into the process, the better. And be open to written as well as verbal feedback, as not everyone wants to voice their views in a room full of people. The quiet ones often have the best ideas.

Diversity of thought is something I have come to understand more recently in my life. The idea is that we all lead differently *and* we all *think* differently. Some of us are naturally late for everything; some of us are naturally early. Some of us see the world as a rosy, happy place; some of us have the complete opposite view. Instead of trying to be right and proving the value of our own opinions, we need to be more accepting of natural differences and try to understand how they work. Diversity of thought means listening to every viewpoint, no matter how different it may be from your own or how difficult it can be to hear the perspective. It also means granting space and time to conversations with people of different backgrounds and perspectives and purposely including them in your problem-solving process. Diversity of thought means being open to your own viewpoint being wrong or, at least, not what's needed in the moment. It takes a lot of confidence and bravery to yield your perspective to another, but those people in the workplace who can truly embrace diversity of thought are the gems among us. Try to grow your own superhero ability in this space to support them. The world isn't shaped by those who quietly follow the rules, but by those who come up with new ideas, speak up even when they're afraid and find ways to accomplish the impossible. We may never fly like Ironman or have the perfect comeback like Peggy Carter, but with confidence, openness and a willingness to be our own superhero, we can manage anything.

TAKE ACTION

We've all heard the saying, not all superheroes wear capes, and this is especially true in the workplace. When we go beyond what's expected and give a little more than the norm, we have the ability to change someone's world in that moment. Think about a recent situation when someone stepped out of their comfort zone and did something that made them a "superhero" among us. Write a quick thank-you note expressing your gratitude.

TAKE ACTION

If you are lucky enough to be able to have kids and want to have kids, do your research! My parenting journey got off to a rocky start due to health and financial concerns, so I recommend everyone think about money beforehand. How much would you need to survive for 3 to 12months off work? Map out different situations and do a cost analysis of all the options. Then save, save, save! Also have a think about which partner is better suited to being at home. Your situation may look different than the norm but do what is right for your family at that point in time.

Chapter 3

How to Be Fearless

"Life is a rollercoaster, you just gotta ride it."

-- Irish singer/songwriter Ronan Keating

Life comes at us with ups and down, twists and turns that rival even the most thrilling rollercoaster rides. Life comes at us with ups and down, twists and turns that rival even the most thrilling rollercoaster rides. I'm sure we've all had the experience of getting a big win and feeling on top of the world -- only to have a crushing disappointment or personal setback bring us back down to earth. In the course of my career, I can think of more than a few situations in which life threw me for a loop. And while I could let the threat of the inevitable drop scare me from reaching too high, I instead choose to hold on, embrace the fear and enjoy the ride. This doesn't mean I'm fearless -- far from it. Instead it means I've found a way to manage my emotions, understand my fears -- and just do it -- whatever it may be -- anyway.

One way I've become "fearless" is by recognising fear as a good thing and seeing the value in being scared of the unknown. Our body's made to function differently when something scares us. Fear starts in the brain, then triggers a reaction involving many of our body's systems. Our amygdala activates the motor functions involved in our fight or flight response, and hormones, such as epinephrine and cortisol, get released. The brain becomes more alert, our pupils dilate and our breath quickens. Our vital organs flood with oxygen and nutrients and our muscles get a rush of blood, all preparing us to react. It's an amazing sequence of processes that take effect at a moment's notice, turning our body into a machine ready to defend itself from an oncoming attack -- or run away at full speed. We've all experienced the flush that comes from a moment of severe fear, and I've certainly felt variations of that at times in my career. It's happened when I've come across a serious mistake I made that I knew would set back a project, or when I was confronted by a client with a major issue. The good news is, that fear response that floods us with hormones and energy can help us be really productive when thinking through a fix. The key is in learning to recognise the feeling of fear -- and then stopping to think What is a logical and rational way forward, you want to ask yourself, instead of running out the door never to return. In moments of fear, I try to slow my thinking with some deep breaths and then rationalise through the steps that could be taken in the moment to correct the problem. I consciously choose to act with intention, rather than react out of fear. And once I'm in control of my feelings, I have the ability to evaluate different scenarios and almost see the situation unfolding before I act. I've learned to harness my body's natural response and channel that extra burst of energy and focus into finding a way forward -- most times, at least. It's far from easy, and even more of a challenge for those who suffer from panic or anxiety disorders that make their bodies hyperreactive to triggers. But when you know your body and your fear response, you can get one step ahead

of the process.

Overcoming fear in my personal life has helped me to

achieve more than I once thought possible. I remember the

fear created by a teacher telling me I was unable to write. The

fear I felt at school when I didn't quite fit in with the other

girls because I loved playing video games. The fear of not

being able to stay in the country I love, Australia was a fear I

lived with for quite a few years. When I decided to migrate

permanently to Australia, I was on a 457 visa, which meant

that my visa was attached to my job. If I didn't do a good

enough job, then that was it I would be sent back to the UK.

When you are on a temporary visa for 2 years for me

personally it was like living on a knife's edge. It got easier

when the company I worked for sponsored me to stay with

Permanent Residency, also paying for half the fees. This

instills a sense of loyalty for a company which has gone

beyond what most companies would do for you. Even then

there was no guarantee. It wasn't until I gained my

Citizenship did I start to feel safe. Over the years this

uncertainty has helped me understand other colleagues who are waiting on the answer to a visa question. I can also understand how frustrating the whole process can be, submitting documents multiple times and chasing emails, phonecalls again and again.

Fear is everywhere and we all have it. New fears come and go, just as new technologies come and go. Over the years, my fears have taught me a lot about myself, including how to listen to the voice inside (the good voice, not the Negative Nancy.) As I've learned to listen to my body and take in information to make the best decision, I've become better at trusting myself to rise above my fears, to be the most "fearless" version of me I can be.

While there are positives that can come from a surge of fear, too many workplaces and workers are more familiar with the downsides. Fear can do more than surge. It can linger and permeate a team or an organisation, creating a culture driven by anxiety and competition, rather than positive energy and encouragement. Individuals working in these environments often become highly critical of their own

-- and others' -- work and they struggle to see the value they bring to an organisation. Eventually, as their disillusionment and unhappiness grows, they begin to question the value the organisation brings to them. This culture of fear pits people against one another -- and even against themselves. One way leaders can unknowingly add to the ambiance is by getting hung up on key performance indicators (KPIs), sales targets and quotas. Periods of turmoil in the team and in the organisation can also heighten the feeling. This environment causes employees to wonder, could I be next? The next to lose out on a promotion or raise? The next to be made redundant. That uncertainty is not a nice feeling to have, nor is the sense that someone is looking over your shoulder, wondering, "Have they worked hard enough? Have they delivered enough? Did they sacrifice their weekend to work?" This type of fear can be used to make sure a team gets the job done -- but employees will not dare to step out of line to get the job done. They will only partially tap into what they're capable of achieving, and the project and organisaton will fall far short of its full potential. An organisation ruled by fear does not inspire people to be thinkers and innovators, two skills needed to succeed in business today and in the future. Instead it creates employees who do what they are told -- and will leave the second they get a better offer.

Different styles of leadership create different levels of fear in the workplace. Autocratic probably best describes the scenario explained above, in which employees worry that any misstep could lead to major consequences. A democratic style, which allows people to be part of the decision-making process, tends to have a lower level of fear and creates a sense of loyalty. Laissez-faire, the choice to let employees take the lead and do what they want, is beneficial in creating trust between leaders and employees, but it lacks a way to hold employees accountable for their output and development. Strategic leadership ensures that the needs of the organisation are integrated with the needs of the individual. In this environment, the leader protects the team and looks after them at all times. These leaders tend to be selfless, driven by a desire to serve others. They will put their teams' needs before their own. Transformational leadership is about setting targets and ensuring individuals grow into more challenging roles and responsibilities. This can be a great way to lead, or an overwhelming one if your employees don't have a natural growth mindset. As a future leader, you will likely find yourself blending these styles and leaning on one or another variation at different points in your career. You will also find yourself being led by many different leaders with different approaches. Being aware of how your leader treats his or her team is crucial. Being aware of how you treat your team is even more crucial, especially as you work to build the trust required to be an influential leader. Think of how you treat your team as a reflection of your own character. A leader who offers up their own role to save two or more members of the team is a rare soul. A leader who believes in training someone else to step into their role is perhaps even more rare -- and the type of leader I aim to be.

I believe that, whatever their style, leaders should lead with heart and with empathy, two powerful antidotes to fear. These traits allow us to make the hard decisions that are right for our organisations, our team and ourselves -- and let us sleep easy at night knowing we didn't compromise any of our values in the process. Compassion and empathy builds trust among team members and relationships that can last a lifetime. And they create organisations that do good -- not just in terms of the bottomline, but for the world around us. With the right understanding of fear and a compassionate commitment to the people around them, tomorrow's leaders can implement amazing changes and inspire great things from those lucky enough to be led by them.

TAKE ACTION

Can you remember a time that you acted out of fear? What was the scenario, and what was the outcome? Write down what you think could have happened if you had been more aware of your emotions in the moment.

TAKE ACTION

Think about a leader you've had in the workplace. How would you describe his or her style? How would you describe the environment he or she created? Note a few things you would do the same and differently -- and why.

TAKE ACTION

Think about this scenario and the consequences of the actions that could be taken. There is never one action there is always a series of actions that could be taken. Which actions do you think will achieve the best results to the situation.

Which actions will ensure that all parties in the scenario are made to feel ok. Can you imagine all of the scenarios and how they pan out? Which one would be the most positive for you in your current situation? Can you visualise this happening again and again before acting.

Chapter 4

On Trust and Instinct

"All it takes is faith and trust."
-- Walt Disney's Peter Pan

How would you describe a feeling of trust? A comfortableness with another person? A sense of friendship or camaraderie? Maybe to you trust means that you can depend on a person, believe what they tell you and know they will show up for you when you need them most. Trust is essential in all good relationships, but it can be difficult to define. After all, it's more of a feeling or an instinct than something concrete we can see and measure. The development of trust between individuals often comes with time, hard work and sometimes great adversity. Teams working against incredibly hard challenges or in situations of emotional turmoil can build huge amounts of trust with each other. (The military is one of the best examples of this.) But the workplace doesn't have to simulate a war zone to foster strong teams. I think of good teams as a strong family unit. The relationships last well beyond the end of the project, and the support and friendship that develop become highly prized. A good leader can recognise the need to create these types of partnerships in his or her team and figure out how to encourage their growth. And in the best scenario, the positive energy one team puts off can ripple throughout the organisation and change the workplace as a whole.

When it comes to figuring out who and what we trust, most of us depend on our instinct. This unconscious radar helps us makes sense of a bunch of clues and cues to give us an initial feeling about a person or situation. It's not always accurate, but it has been known to help people find their soulmate (love at first sight, anyone?) and has also been credited with many successful business decisions. (Ever "have a hunch?") Sometimes, the first level of trust we need to build is with ourselves and this instinct tool. In my experience, trusting my instinct starts with listening to myself and my feelings. Much like being "fearless" starts with recognising the feeling of fear, trusting my instinct starts with learning how this great, natural tool communicates with me. I pay attention to questions I seem to be continually asking myself and notice where my interests have been lately. I "feel out" a situation by really paying attention to the feelings I'm experiencing. Am I nervous in the presence of a teammate, or excited to see them? Do I feel reluctant to share my ideas with a certain person? Why? While I aim to listen to my instinct, I'm also aware that the body's unconscious judgments aren't always correct. I try to err on the side of trust, going into relationships with openness and giving everyone the benefit of the doubt while still being aware of my initial reaction to the person or situation. I then seek to understand why someone is behaving the way they are. Is it a personality trait, or a reaction to the way I am making them feel?

In order to perform well, a team needs to have trust, purpose and a common focus guiding their work. This has, perhaps, become trickier as the workplace has opened up to so many types of people from different cultures, backgrounds and places around the world. The fact that teams often work remotely in today's business world, with members dialing in from Bangalore and Berlin to share ideas on a conference call, adds another layer of difficulty. We might not always feel a natural, easy connection to the people we work with. But we need to strive to achieve this. Some of the best teams I have been a part of are, at their core, *fun*.

Going to work feels more like going to spend time with extended family, and teammates often do things that go beyond work colleagues. They make a cake for a birthday, cook meals for a team member when a newborn arrives, check to see if everyone is OK before they leave for a Friday. Above all they pull together when a member of the team is low or going through times of hardship or illness. They live by the motto, "We look after our own." If, as leaders, we can learn to do this in our teams and inspire others to do the same, our business and communities will be a better place to be.

How do we know when our teams and our leadership is working at its best? I believe it happens when we reach a state of being I'll refer to as spiritual intelligence. This occurs when our emotional Intelligence and our rational intelligence fuses to create something of a higher order, an ability to know and understand what is happening intuitively with other team members. This can lead to moments of déjà vu, moments of knowing what the other person has to say. It happens when we have short-hand conversations with each other that an outsider would struggle to understand. I've experienced moments when team members turn up to the same social occasions, send an email to each other at the same time, ring each other at the same time, or pick up the phone at that exact moment. We've texted each other or bumped into each other and said, "I was just thinking about you." These moments let you know your team has become a cohesive unit and take on any task on their to-do list. They can deliver far beyond what they thought possible. They can invent, think differently and blend ideas in a way that could change the project for the better, or maybe even change the world. In order to reach this state, a team needs to collaborate at a level that works on trust, free of competition and threatening feelings. Above all, they need to be honest with each other and respect each other's abilities and views.

Walt Disney was known for creating strong teams with a strong mission, and his recipe for success continues with the Disney company, even though he is no longer around to see it. "First, think. Second, believe. Third, dream. And finally, dare," said Mr. Disney, an amazing businessman who achieved things most of us can only dream of. His movies and the Disney films that followed his death often speak to issues of trust and instinct. Think about Pinocchio and the trials he had before learning to listen to his conscience. In more recent years, there's been great female characters, such as Mulan and Moana, who trust their instincts to break out of social norms and follow their destinies. Trust is also

something that's present in the movie *Frozen*. The wonderful

female characters must learn to trust themselves and each other,

with the emotional Anna learning a tough lesson about trust

misplaced and Elsa opening up to her feelings to trust herself and

others. Together they become the perfect team, but they have to

work to achieve it.

Another awesome Disney character -- one of mine and my

children's favourites -- is an outspoken little fairy named

Tinkerbell. In the original film of *Peter Pan*, Tinkerbell is cute as can

be but has a feisty, fiery side. She's sometimes selfish, sometimes

annoying but always fiercely loyal to Peter Pan and those she loves.

By the end of the film, she even overcomes her initial instinct and

selfish side to trust and love one who proves herself worthy,

Wendy Darling. In the newer films, in which Tinkerbell and a band

of friendly fairies star, Tink is known for her skills in "tinkering,"

building and problem solving. She's a good friend, loyal to her

core, resilient if a bit stubborn, and often helps to save the day. The

way Tinkerbell connects with others has me thinking about the way

I connect and interact with different people. Sometimes I seek out

the Tinkers in this world, people like me who like to build, create or reuse. Sometimes I almost see a spark in people, perhaps a spark of creativity, as if they're "switched on behind the eyes." The biggest spark I will ever remember seeing was when I met my partner. I felt a need to find out more and figure out what was behind his light. An instant connection like this can feel almost as if fairy dust has been sprinkled over the relationship, as if you're floating off to Neverland. It's important to make sure these kinds of sparks don't get lost, to reach out and reconnect if the spark starts to waiver. While instincts and unconscious connections play a big role in the workplace, conscious actions taken when we are aware of our influence on others can be even more meaningful. SpaceX's Gwynne Shotwell comes to mind. She is the COO to CEO Elon Musk, a pair of leaders that are changing the world one idea at a time. Leadership often looks like this, with one person creating the vision and another bringing it to life. Again, this reminds me of Peter Pan and Tinkerbell. Some entrepreneurs resemble Peter Pan with their childlike wish to do the impossible. They need a Tinkerbell who can speak out with honesty, tinker with the ideas and add the spark to make them fly.

While we've talked a lot about building trust with our teams and in the workplace, there's another layer of trust that's perhaps even more pressing for businesses to develop -- that between the client and the organisation. Trust with clients can be built in the same ways trust arises between team members. An organisation can aim to create a "family" relationship with a customer by going above the expected in-service delivery, truly caring about how a client feels and having some fun along the way. A close customer relationship -- like a close team relationship -- can lead to the same "spiritual intelligence" I discussed above. A company can come to recognise the client's needs, perhaps even before the client knows what he or she wants. In the digital age, data makes this type of intelligence a real possibility. We can glean many insights from trends and anomalies in the customer data we collect. But collecting and analysing data also puts a company at risk for losing customer trust through a very real threat -- data breach. If the data held by a business is not looked after with great care, then the loyalty a customer feels toward the organisation can be destroyed. Data is powerful in today's world, and the strategic insights that can be gleaned from data client data make it invaluable. An organisation has to be brave enough to understand the data, ask the right questions and apply the right knowledge to data insights, but it must also be loyal to customers and vigilant in protecting the intellectual property.

Trust is an incredible tool for building great relationships in teams and with the client, but it doesn't happen on its own. It takes work, personal awareness and a bit of faith to build trust with the people we work with and the people we serve. Oh, and something else I forgot. Dust, pixie dust -- that little bit of magic that can take a normal idea and make it fly. "All the world is made of faith and trust and pixie dust," said J.M. Barrie's Peter Pan, and I couldn't agree more.

TAKE ACTION

Are your instincts trying to tell you something? Take note of any questions or ideas that have been going through your mind on repeat lately. Jot them down and see if you can determine a pattern. Your subconscious might just be trying to share an insight with you.

TAKE ACTION

Identifying with a character from a movie or book can be a great way to understand yourself better and even encourage yourself to step out of your comfort zone. I feel a likeness to Peter Pan's Tinkerbell, which can help me when I need to make a tough decision. ("What would Tinkerbell do?" I can ask myself.) Think about a character you relate to and write down the characteristics you admire or would like to emulate in the workplace.

Chapter 5

Learning From Your Mistakes

"I can accept failure, everyone fails at something, but i can't accept not trying."

-- Athlete Michael Jordan

You've missed the mark, backed the wrong horse, gone pear-shaped and caused a Charlie Foxtrot. In short, you messed up. You made a mistake. You *failed*. Oof. That's not a good feeling, is it? Now own it, accept it, apologise and move on. The English language has a ton of euphemisms for failure, about as many as we have for death. And I have to think that's because we humans enjoy the feeling of failure just about as much as we do kicking the bucket. We might even prefer giving up the ghost to going up in smoke. Feeling bad about a mistake is a natural reaction, but the digital age we live in requires a very different mindset -- one

43

focused on embracing failure and trying again ... and again and again. It can be difficult to change your thought patterns and adopt this so-called growth mindset, but it's absolutely essential for leaders of today and tomorrow.

The term growth mindset has been trending in recent years, as everyone from educators to entrepreneurs preach on the importance of taking risks and learning from mistakes. Stanford psychologist Carol S. Dweck is often credited with turning grit into a sought-after personality trait. Her research showed that while some students rebounded quickly from setbacks, others had difficulty mustering the courage to give it another go. She coined the term fixed mindset to describe the thinking of people who tend to give up, and growth mindset for people who try, try again, believing that if they work harder they can achieve their goals. Leading with a growth mindset means believing that from every experience -- the good, the bad and the ugly -- we can learn something. It means being aware of what you're good at and what you need to learn. And it means sharing these skills with your team. Your list of strengths and growth areas should never be one-sided as it's crucial to know both. And we must remember, all weaknesses can also be strengths -- and vice versa. An example of this is one of my core strengths -- honesty. When asked what I think, I cannot lie. If I fundamentally believe that an approach is not the right way, then I feel the need to share my version of the truth. This can have drawbacks, as some people do not like hearing my truth. And there have been times when people have been hurt by my honesty. I've learned that when our world is black and white and not shades of grey, our perception of right and wrong can get warped. I must remember that I *can* be wrong, my version of the truth is not the only one. I've learned through experience that what I see as honesty can actually be hurtful at times, especially if the timing is not right or the words not chosen carefully. Honesty has become a growth area for me, as what I once identified as a strength has also revealed some weaknesses.

The first time you see someone crumble from feedback you have given or because you had to end a contract due to poor performance, you realise the challenging human side that comes with leadership. These experiences have led me to focus on a growth mindset with my team as well as myself. I've made it my goal to find a fitting role for a person experiencing failure, a role within our organisation that better fits his or her strengths and needs. I've learned to never think that because someone has failed at one task they cannot learn a new skill or succeed in a different opportunity. Instead I've tried to find a role that better suits them. I give them time to think about what they want and where they want to go next. I help them determine their strengths, their weaknesses and believe in them along the way. By taking this growth-centered approach, based on honest and sometimes difficult conversations, I've found I can help underperforming team members experience success and even become a leader in the future.

In times of failure, a leader must be very aware of how feedback is offered -- is it meant to do harm, or to help the situation? Feedback given in anger or loaded with blame is one of the worst forms that can be given. It can damage team spirit and lead to unintentional harm. Unhealthy teams -- those focused on competing with each other and getting ahead as individuals -- often use this type of feedback to the detriment of everyone involved. And the product or delivery suffers as well. In the field of technology, honesty is an amazing gift for someone to have in a cybersecurity role or when interpreting the law. However, when you are dealing with changing people's perceptions about a situation, it is often best to plant the seeds of thought and help people realise what needs to be undertaken themselves. Sometimes the speed at which you reveal your reality to others (your team or your clients) needs to be slower so you know they're keeping up with you on the journey of change. Too much too soon can scare people away. You must introduce people to the possibilities ahead and lay out clear steps for how you plan to get there. Collateral damage happens, however, and sometimes it's even necessary for change to occur. It takes a brave soul to stand up and say I don't agree with the direction we are currently heading, in our team relationships, our products or our technologies. Master the skill of getting people to buy into your reality and your leadership will flourish. Also always pause to listen and ensure you're ready to learn from those willing to share.

In my technology career, I've watched patiently as clients moved up the realisation ladder to figure out a particular solution or technology was required. Eventually, they started to believe they could make a change and that the time was right to do so. New technologies have a tipping point, a moment when the business case for adoption eventually stacks up. And in some cases, the right answer to the question of adoption is not "no" but "not yet." Over time, as technology gets cheaper and is tested and proven, the client is more likely to see the advantages. And when organisations see others around them start to move on a technology, they believe they should too -- and quickly. The rest will follow and no longer be laggards. Organisations that plan strategically, creating a roadmap for future technology changes, are the ones best placed for success, no matter how big or small they are. These organisations can quickly learn from mistakes and pull forward a deployment, if needed, in response to an accident, incident or a need to innovate. Of course, organisations also learn from doing too much too soon, from going back to the drawing board and trying something different. There's always failure and learning in the technology space, it's just a part of our world. And the sooner we and our clients accept that and adopt a growth mindset around it, the better off we'll all be.

Over the next 20 years, the growth mindset will continue to be important, but I think those that teach others how to change the way they think will find the most success. The youth of today -- those people who will lead our organisations in 2040 and beyond -- will need to be brave and bold. They will need to inspire people to think beyond what they see today and the roles they fill. They will need to learn and relearn in order to transform our world. The leaders of tomorrow will need developed skills but also the ability

to develop brand-new skills, as they create and adopt technologies

that don't even exist today. Think about the way in which a new

app is deployed to your phone today, or the way in which a new

virtual machine is launched in the cloud. We take for granted that

the task can be undertaken in less than 30 minutes, if you have the

right access and skills for the job. But in the future, an artificial

intelligence will be able to set up a new virtual machine (VM) for

us. Or perhaps we will deploy a bot to learn the role and let us

know how the process can be optimised -- and then the bot will

take over the job altogether. These technological advancements will

mean our future leaders will need to know how to work with ever-

evolving types of intelligences, something that should come

naturally to a generation that has grown up communicating

between the digital and analogue worlds. These future leaders will

have the skills to navigate the next wave of computer-to-human

interfaces. They will be the innovators that make the sci-fi of today

the technology of tomorrow.

Another part of my leadership journey and a lesson I have

learnt over the years is becoming aware of my own unconscious

biases. I was very lucky to work with a graduate a few years ago who taught me so much. Yes I mentored her, but she also

mentored me. As part of a research project we were working on,

we built an AI that would consider the biases of humans in the

hiring process. In order to do this, we needed to understand the

unconscious biases that different people have. The graduate

introduced me to more than 100 different prejudices and computer

software that can help people identify their own. Slowly but surely,

I became aware of my own unconscious biases. One was a bias

towards those who don't have an online profile, a digital footprint.

If a candidate didn't have an online profile, then I was less likely to

be aware of or interested in them. Even worse, I was initially

disgusted to learn people had disconnected intentionally from

social media. I had no idea my biases in this area were so strong.

Once I became aware of them, I was able to understand the

different perspectives and learn why someone would choose to not

be online. With time and patience, I learned that I do not have to

agree with a perspective, but I can still respect people with differing

views and understand their reasons. Giving people the time to

explain a viewpoint has allowed me to become a better leader.

Another example of my own growth perspective is

extremely personal. When I was younger, I never fit in with the other girls. I was always geeky, loved computer games, maths, numbers and seemed to get along better with boys than girls. I chose not to do maths in class, mainly because I did not want to face any more ridicule than I was already receiving for my bottle-top glasses. When I first came into the technology industry, I was warned about working in the space. I was told it wasn't the place for a woman to be and that many people who worked in the industry ended up having mental breakdowns. There were no such things as Women In Leadership or Women In Technology groups and awards. The chances we were given came about because we competed equally against everyone else; we had no choice but to do so. I have to admit there have been times when I wondered what I was doing in this industry, but my setbacks and failures only pushed me forward. When I turned around 26, I helped to build a group for Women in Technology at an engineering firm where I worked with the encouragement of my female manager. Though I still believed that the best person for the job -- no matter their race, diversity or gender -- should do the job, I became a little less naive. I realised that people tended to hire people like themselves, and in technology that often-meant men. Women had to compete to get what they wanted, and in the beginning of my career, many of us believed there was not enough space for all women to be at the top. Even I believed this! Now I realise that when women work together and help each other, more can be achieved. I have seen first-hand that there's room for all of us. I have also seen the benefits of helping other females achieve and thrive in their chosen careers. This year, I attended my first all-female awards ceremony, Women In Technology Western Australia. I won an award for Social Media, and I felt so proud standing on the stage with my husband smiling up at me. My mindset has grown so much over the years that I can look at who I am and how I've succeeded without those biases that once drove me. I am a female. I am in technology, and I am a leader. This is who I am, and I am happy being me. And I will help everyone of all shapes, colours and sizes that follow me in my footsteps.

In order to grow and learn in the 4.0 world that will develop over the next 20 years, we all need to adopt a growth mindset, one that accepts and learns from failure and feedback. We also need a futuristic mindset, a belief that anything is possible. Some might say society had this mindset in the 1960s when we raced to send a man to the moon. How far we have come since then is nothing compared to how far we will go in the years ahead. Over the next 20 years our minds will confront changes we can't comprehend, and our world may be unrecognisable at times. We will encounter failure and setbacks, that's a given. But with a mindset focused on learning and growth, we can find our way to success -- and inspire our teams to do so, too.

TAKE ACTION

Think of a recent failure or setback you experienced. How did you feel in the moment? If it was a negative feeling, what did you do to deal with those emotions? Take a moment now to write down what you learned from the failure and focus on the positive side of the event. When your next setback occurs, think about this exercise and try to shift your thinking to a growth mindset.

TAKE ACTION

Encourage your teams to invent and create. Encourage your team members to read and watch Sc-fi and not be afraid to broaden their horizons about what they learn. Learning can come in many forms, including the activities we do for fun.

Chapter 6

Thinking of Others

"Unless we think of others and do something for them, we miss one of the greatest sources of happiness."

-- Former Stanford President Ray L. Wilbur

We've all had this experience in the workplace: We're seated at a table or logged into a conference call and, when asked our opinion about something, the answer comes out less than eloquently or we forget to come off mute. We stumble over a few words or say something in a way that might not make complete sense. The meeting moves on, but we feel a flush of embarrassment that lasts the rest of the day, maybe longer. We replay the cringe-worthy moment over and over in our minds. "What are they thinking about me after that terrible performance?," we worry. "I bet they're talking amongst themselves about how I'm incapable, unqualified and should be kicked off this project as soon as possible!" Ever had these thoughts or something like them? I know I have. But here's the thing: Those colleagues at that meeting? They probably weren't thinking about you at all. They were probably too busy worrying about their own perceived gaffes and missteps or perhaps what they were going to have for dinner.

Research has shown that humans primarily think and talk about themselves. One of the main reasons we even talk to each other at all is to share information about ourselves and our perception of the world. And when we're alone with our thoughts, we're mostly thinking about yours truly too. Scientists have located an area in the brain that switches on like a screensaver when we're not engaged in more demanding tasks. That screensaver happens to be the same area of the brain that lights up when we think about -- you guessed it -- us! There's a fascinating body of research looking at why our brains may have evolved this way, a survival instinct at its core. But being aware of this truth does two things: It frees us a bit from concerns that everyone is judging us in the same harsh way we judge ourselves. That's a good thing. But it also makes us aware of our own bias to focus only on ourselves and our own needs. Thinking about others can be hard work, especially when those "others" seem, on the face of things, to be so different from us. People communicate in different ways, share and accept praise in different ways. People think in different ways and bring a lifetime of experiences to their own ideas and opinions. There are many people in the world who, on first glance, we might judge harshly to be "weird" or "strange." But when we start to pull back the layers of our own biases, we're able to spark a great connection.

The key to embracing differences is understanding. If we can understand what we think and why, then we can understand where our bias is coming from and learn from differences. We can even learn to use others' differences to complement the way we work. In technology, we can create better solutions and products when we add differences, or market differentiators. And the greater diversity we have in our teams, the better the product we deliver. Diversity of thought is something I touched on a bit in Chapter 2 and will speak to again here. A leader that embraces the belief "that more ideas are better" listens to every viewpoint, no matter how different from their own. They encourage the problem solvers and ask questions that lead to new answers. They identify others who can make a difference and encourage them to speak up, show up, step up, draw up and present what they think. The world wasn't created by those who followed the rules; the world was created by those who came up with new ideas, who continued to try when they were told something was impossible. Leaders focused on others will make space for these world-changers to thrive in their organizations.

One way we can get out of our own heads and focus on others is in the expression of gratitude. An authentic expression of thanks can make a lasting impression on a team member or colleague. The genuine connection you make when you express thanks for the sharing of a good idea, a kind word on a tough day has the ability to change the world for a moment, to make someone feel great and, perhaps, *be* great. Another way to focus on those around you is to take time to communicate with them. Send a simple email or text to check if someone is OK, share a news article about something you know they're interested in. Volunteering or committing yourself to a charity is another wonderful way to expand your focus and show that you can think of others above yourself. And though not all of us are cut out for the military, it's the highest example of service and sacrifice I can think of. Some of the greatest leaders I've met came from different military backgrounds. They are strong servant-leaders who look after others before themselves.

In the technology industry, a key example of "thinking of others" happened in 2019 when Microsoft's CEO supported the move to "make peace with Linux, saying that it will allow more than 2,600 other companies, including long-time rivals like Google and IBM, to use the technology behind 60,000 Microsoft patents." For me, this was an example of a company refreshed, shifting its focus back to its original mission: to change the world. Back in the 1980s, Microsoft aimed to "put a computer on every desk and in every home," which seemed far-fetched back then. Though the company may have lost sight of its person-focused mission over the years, I feel it's returned to that spirit by ensuring technology is available to all. Microsoft likely realises that we, as consumers, no longer want our corporate giants to focus only on the bottom line and performing well financially. We also expect them to do the right thing socially and culturally, and to express gratitude to their customers. We expect them to go beyond the service or product delivery and do the extraordinary to help the community in which they belong. I think Microsoft will inspire others by this example, a game changer that will allow any Internet of Things device developed through the Linux platform to be understood in detail with patent designs and descriptions available to others. This will, in turn, inspire ideas and empower others to create something new from what came before. The act may also inspire other technology organisations to open their intellectual property up to more entities. I wonder what organisation will be next, and what we will all gain from it?

One area of technology that definitely requires a "thinking outside ourselves" attitude is Artificial Intelligence. If used in the right way, AI can help humans do more and become more productive. It can help us with big problems and questions, such as how to cure cancer, find a vaccine for Covid-19, solve climate change or feed billions of people around the world more efficiently. At the same time, AI has the capacity to harm. If we look at many of the data breaches and other security concerns, we have seen over the past five years, we'll see AI being used in a manner that doesn't align with our moral values. AI does not have a built-in obligation to do good; we humans have to give it that. Unfortunately, not all people or companies are willing to think of others when designing and utilising AI. Recently, companies such as Microsoft, Apple, Amazon and Google took a stand against the use of AI for devious purposes. They said "no" to forcing and breaking security in order to allow access to encrypted machines or data centres in other jurisdictions of the world. These decisions take courage. But one right move now won't protect us in the years to come. We will need to stay vigilant to make sure our technology protects and benefits us, that our companies are thinking of us, and not just the bottom line, when they develop their products.

It's human nature to focus on ourselves, our own worries and perspectives, but we can achieve so much more when we get out of our own head and start thinking about those around us. As the Buddhist Monk Shantideva wisely realized in the 8th Century, "The source of all happiness lies in thinking of others." Leaders who think of others, who welcome diversity of thought and practice moments of authentic gratitude, build stronger teams, create better products and deliver improved services. And companies that can focus on the customer, the competitor and the world beyond, instead of their own bottom line, can change the world.

TAKE ACTION

Think of a time you got stuck in your own head at work or in a personal relationship. What's something you could have done differently to refocus your attention away from yourself and toward another person? Would it have helped you feel better about the situation?

Chapter 7

On Innovating and Inventing

"There's a way to do it better. Find it."

-- American Inventor Thomas Edison

Doing what has never been done. Going where no man has gone. Pushing the boundaries and rocking the boat. Many of us don't think twice when we use phrases like these to describe being on the "leading" or even "bleeding" edge of a situation. But just because we use the language doesn't mean we have what it takes to be an innovator in business today. Over the course of my career, I've identified several groups of people who approach innovation in different ways and create varying levels of impact. Understanding these types and their motivations can help you make a case for innovation.

First is the **Edge Follower.** This is someone who wants to be on the leading edge because, well, it's edgy. It's trendy. It's good for the ego. These people or companies tend to follow the lead of a true innovator but aren't the ones setting the trends. They tend to be financial services companies, large companies that don't want to take too big of a big risk. They usually wait to buy other companies or adopt changes once they know the technology has been proven. With a little guidance and a boost of confidence, an Edge Follower can become a trend-setter of his or her own.

Next is the **Edge Soloist**. These individuals go it alone, for the time being. These people have not yet developed their ideas enough to share with the world but they may have one or two trusted confidantes to discuss their goals with. Scared by the unknown, Edge Soloists must work to overcome their fears and advance to the next level of innovation. They can often be mistaken as a follower and easily misunderstood, but, in time, they can develop the confidence they need to share their vision. These companies are the ones which are often bought up, the money is too great for them to turn down, take companies like Linkedin, UIpath and Twitter. To name a few, they were small companies to begin with bought up by the big giants, they were great ideas which were fostered and looked after, we do not necessarily know straight away who created them.

The next level is **Edge Visionaries.** These people push the boundaries and see the future before the rest of us can. They see innovation as revolutionary and often thrive in extremely collaborative environments. They worry more about realising a vision than having that vision adopted by the wider market. These tend to be people like Elon Musk, they thrive in the moments where they can be on their own, this is when they are a t their best, they can see things before anyone else can.

The final and highest level of innovator I've identified is an **Edge Exponential Opportunist.** These individuals are on the leading edge because they see a pragmatic opportunity no one has exploited yet -- and they know how to take the lead to make it happen. They see innovation as a necessity and invention as evolution, a survival trait they need to succeed. They achieve great things and lead their organisations to new heights. Edge Exponential Opportunists tend to be collaborative and encouraging in their work. They often spot nascent skills in others and help them develop their talents.

It's fascinating to think about innovation as both a built-in personality trait and a skill we can develop, a ladder we can climb. We also move on this scale, depending on what we are doing in life and where we are in our careers. I also thing that depends on the amount of risk that we are willing to take. When I first started writing this book I was an Edge Follower, to scared to go it alone, with the protection of a big four, to voice my view point and be heard in a different manner. When I found my voice, I could no longer be silenced, I became an Edge Soloist, Now I am certainly an edge opportunist, not quite an edge visionary just yet, although one day I would certainly like to be. I am leading to make this book happen, I am taking a risk doing it, however I know If I don't do it then the words will sit inside me like a defiant and naught fairy waiting to escape. Who knows people might enjoy this book, they might not. Either way I give people the opportunity to discuss, interpret and misinterpret what I have said.

I give the gift of this book so that others can add to the ideas to the notes, to the thoughts so others can develop, ideate and elaborate.

Leaders in the technology space especially need to keep up their skills, swap ideas with like-minded people and maintain their excitement for an industry that is always in flux and racing toward new horizons. If we lose our passion and love for new technologies, we can potentially drop to the category of Edge Follower. The best leaders will cultivate an Exponential Opportunistic mindset, one that loves being on the edge and taking others with them. These individuals have realised that when great minds come together to create something new and extraordinary, the edge is a less lonely and much more fun place to be. And companies have realised the value these leaders bring to their organisations. With their passion and ability, Exponential Opportunistic leaders work to transform their organisations again and again to meet the demands of today and tomorrow.

I have seen this firsthand on numerous occasions, in companies forming partnerships with the company I happened to be working with at the time, the first was in 2009, when I got to create a new company from partnerships, we had formed with a small organisation. I have seen this happen again and again this is when I have seen amazing innovations come together. Over the years I have been lucky enough to work with Google, AWS and Microsoft to create amazing technological advancements for clients. I have also seen clients do this themselves, sometimes successfully and sometimes not so much with parties going their own ways. I think the key to this is a trusted partnership, is the trust is not their then innovation cannot occur. The teams have to blend together perfectly and also trust each other. Without trust there will not be the opportunity to innovate.

TAKE ACTION

If you're like me, you're bursting with ideas. Personally, I need more than five projects on the go at once to keep me from being trouble! I enjoy changing the world. I enjoy making a difference. If you are one of these people too then power to you! But remember, you can't tackle it all at once. Take a few minutes now to write down ideas for future projects you could take on at home, at work and beyond. They may not be ready to come to fruition yet, but their time will come. And the 5% that really do well could be life changing.

Chapter 8

Creating Your Balance

"The best and safest thing is to keep a balance in your life,
acknowledge the great powers around us and in us.
If you can do that, and live that way, you are really a wise
man."

-- Greek Playwright Euripides

Life can feel like walking a tightrope. We take careful steps
to stay centered and in control of everything going on around us,
but one misstep in our career or stumble in our family life can send
us teetering over the edge. It's tough ensuring we have just the
right amount of *everything* in our lives. That we're both active and
rested, have social lives and moments of solitude. That we spend
quality time with our children and partner and quality time with
ourselves to recharge. We have to make sure we're not too
introverted or too outgoing, too colourful or too plain, too busy or
too relaxed. It's a constant act of give and take, trial and error as we

figure out what we need to stay balanced and upright.

One thing I've found that helps keep me balanced is doing what I love. When we take the time to do what we love — to write, to be with our children, to learn, to develop, to mentor, time for a hug from a loved one — time can pass without us realising it. Life can feel calm and centered, even if just in that moment. When I'm not doing something that really engages me, I find the world can pass by unplanned and undecided. I feel off-kilter and worry about the value I'm bringing to those around me. When I feel like this, I purposefully stop and re-centre my actions. I take a deep breath, take a walk outside or go up and down floors of my office to say hello to people and gauge how they're doing. It's a centering experience connecting with colleagues face to face — and it adds to my Garmin activity as well!

Something else that works for me in my constant quest to maintain balance is asking for help when I feel lost or unsure about where to go next. I often turn to people in my inner circle for advice and guidance. These individuals know me best, and I can trust they have my best interests at heart. A few of these people are professionals I have met online like Mark C. Crowley, coaches I have worked with before, including Ro Gorrell and Ian Sharpe, and previous managers Frank, Phil, Penny, Mark and Mark. There are a few clients that I also trust to go to for advice and guidance (Cathy, Glynn, Richard, Lance and I could go on for a while). Most of these people have a lot more experience than me and have been there before. I also turn to people younger than me for a different perspective. They keep me on my toes and offer a view that's often different than my own. They can be awesomely honest, which is refreshing. (Carlos, Courtney, Sharon, Nick and Alex, thank you). There are quite a few technical people that I can also depend on to get an opinion about a vendor software or how to solve a complex problem (Rob, Andy, Max, Johannes and JVB). And in my personal life, I can trust my partner, sister, mum, dad and my in-laws to offer thoughtful advice when I need it. Then there's Christine, my editor, who gets to read my raw thoughts before they are eloquently edited so that everyone can understand what I am saying. Reading this paragraph back, I realise how very lucky I am to have an amazing global network of super smart people who would help me at a moment's notice. Thank you, all, for being you and being honest with me!

Sometimes when I'm struggling to find balance or come to a decision, I reach out to people I know will have a different perspective than my own. These real-world interactions can help me think differently, lead my mind in new directions and bring to light new arguments and counterarguments to shape my thinking. Someone of a different generation may offer a new perspective, as will someone from a different country or culture. I often reach out to people I've developed relationships with online for a global perspective on a question or concern. One reason to look beyond our own backyard when asking for advice is to prevent confirmation bias. Confirmation bias is a natural human tendency to see what we want to see in the data, research or opinions we're offered. Technology has made it even easier to fall into this trap. Tools like Facebook and Twitter show us news and information based on things we already liked or engaged with before. Thus the view of the world I take from my Facebook feed is very different than the ones my partner, boss or mother sees. In this day and age, we have to intentionally search out differing points of view -- otherwise we will be at a real disadvantage in making decisions. Sometimes we even have to un-think what we have been taught in the past in order to *expand* our thinking. I still think that Pluto is a planet, as it was drummed into me at high school. Though scientists now say it's too small to be a planet, the small, bluish plutoid was always one of my favourites and I hate to give him up!

Something we may recognise as truth at one point in our lives may change over time, as the rules shift, or new information comes to light. Or we may recognise our own bias toward the situation and push ourselves to shift our perspective for a better understanding.

Perspectives can change over a short period of time, and leaders must always be aware of how they shift and spin. As leaders, we often see allegiances form, storm and explode, like invisible threads of thoughts traveling across a room, a city, a continent or world. One day these threads of knowledge may even span virtual and alternate realities, as we gain the ability to tap into multiple layers of information from, perhaps, multiple dimensions. Leaders who are aware of different truths, based on individual perspectives, experiences and emotions, gain an extra edge. And, as Artificial Intelligence plays an even bigger role in the workplace, leaders may tap into an even higher level of intelligence. I believe the Superpower of tomorrow will be the human ability to think differently, to tap into an emotional intelligence that allows a deeper understanding of our colleagues and clients. Artificial Intelligence can help us reach into these sources of truth, map and layer them onto other pieces of information to present a fuller understanding of the situation. Already we're starting to see that happen. My phone battery "dies" on a day when I have connected to too many things or people. My watch will sometimes do the same. We depend on these tools to provide us with a constant stream of data, to put knowledge at our fingertips and connect us with people around the world. Our digital technology allows us to discover and use "truth" in every decision we make. It makes me wonder at what point this stream will become a type of AI consciousness we can tap into with the simple act of thinking about it.

Social media has given us a glimpse into that future, though the "truth" it presents is not always the reality I know. Platforms like Facebook and Instagram often present a filtered view that I can imagine being questioned by future generations who look back on what we were up to in 2019. The digital archeologists of 2040 might wonder why they see only happy, smiling photographs in the social media artifacts we leave behind. We don't often show the trauma that comes with breaking barriers and doing things differently. (Although the Black Lives Matter protests of 2020 have seemed to change that a bit.) We don't often share with our social media networks how difficult change can be, how difficult *life* can be. If we show the true story behind the photograph and let others know, "life's not always rosy and the journey is often bumpy," I think more of us will be able to deal with uncertainty in a more measured manner. I'm always inspired when I see people being honest on social media, when I see a working mum let others know she has suffered from post-natal depression. When I see a family with a child with autism open up about the struggle to get through tough days. Perhaps it's an entrepreneur lamenting the challenge to get funding for an awesome start-up idea. These human struggles are not often shared front and center on social media, but they are what connects us to each other. They inspire us to keep going during our own challenging moments. And they remind us that life's a great balancing act. We all need a little help to hold steady.

TAKE ACTION

What's an area that feels out of balance in your life today? How is your mind or body letting you know that? Take a minute now to write down some ideas for finding balance in this space. Consider centering activities, such as connecting with a colleague or a friend face to face or spend some time outside. Think about reaching out to a friend, mentor or online connection for a new perspective on the situation. Be aware that everyone goes through periods of struggle, even if your social media feed would have you believe otherwise. Be patient and kind to yourself as you work to find your balance again.

Chapter 9

First Connect, Then Build a Home

"I define connection as the energy that exists between people when they feel seen, heard and valued; when they can give and

Receive without judgment; and when they derive sustenance and strength from the relationship."

-- Researcher and Writer Brene Brown

I live and work in the world's most remote city, Perth in Western Australia. I love Perth, a place I landed temporarily for work years ago and found myself moving back to permanently not long after. I met my partner here, and we raise our two daughters on a farm outside of the city proper. It's a wonderful, beautiful place -- but close to the rest of the world it is not. Thankfully physical distance doesn't matter much today -- in business or in life

74

-- because we're all just a few clicks away from each other. Digital technology and social media in particular have changed our sense of place, our perception of others and expanded the networks that connect us all. My children can dial up Grandma and Grandpa in the UK anytime with apps such as WhatsApp. I connect with colleagues in New Zealand, Washington, D.C., London and all over the world with networking tools like Google Hangouts. As someone who often works, physically, away from clients and colleagues, I very much appreciate the online connections I make. And those grow every day. I started using Facebook in 2007, Linkedin 2008 and Twitter in 2009. At the beginning of February 2018, I had more than 1,400 connections on Linkedin and by October of the next year, that number had almost doubled to 2,640 Linkedin followers. My Twitter account has had similar growth in the years since I started tweeting, and I now have about 1,500 people around the world reading my thoughts and engaging with content I share in that space.

Interconnectedness is something we all crave, no matter if we're an introvert or extrovert, from Perth or Peru, a boomer or a millennial. It's a human desire to connect with like and different minds, to hear others and be heard ourselves. We all want to share our story and share *in* the story of humankind that we're creating together. And social media allows us to do that in ways that were unimaginable just 20 years ago. In the future, I think our desire for connection will take us in even more novel directions. We will connect more with Artificial Intelligences, something some of us have been doing in recent years with Siri and Alexa joining many of our homes. And perhaps in the distant future, we will engage with humans and intelligences on different planets, not just different continents.

If you were to know me only through my online persona, you might have a certain perception of who I am. You might think, for starters, that I never sleep! I seem to post day and night to my social networks, but I have a confession to make on that front. Most of the posts I share have been chosen by me but are scheduled and posted by a tool that selects the best time to reach my clients and connections around the world. Often it isn't me hitting the "share" button numerous times throughout the day or night. It's a robot doing it for me. And I would venture that many of the online personas we engage with everyday are using the same kind of tools. We live in a world that doesn't stop, and time is our most valuable resource. But we all need time to disconnect, re-centre and experience our physical self without digital distraction. I've found a way to balance my online life with my real one. Want to catch the real me on social media? I limit myself to logging on during my commute time. This helps me to focus and have time dedicated to responding to emails and online messages. The rest of the day, I avoid the distraction of social media and focus on the real work I need to get done -- while bots keep my social feed humming along.

Social media has the potential to make life more complicated but it also makes tasks so much simpler. That's something we take for granted now that many of us can barely remember a time without these tools. Keeping in touch with people, hiring people into my teams and also finding out what is happening in the industry, I can do all of that from behind my screen in Perth, on my commuter train, wherever I have a connection. Digital technology and social media gives us the tools and capability to listen and be aware of others, a valuable gift. And when people feel valued and listened to, connection naturally follows. My own social media journey went from connecting to different people around the world, to, in 2015, learning how to use these tools to my advantage to grow my career, share my writing and spread my viewpoints. A very wise and wonderful digital strategist at my company, HJ Higgins, taught me how to deal gracefully with online feedback, to say "thank you" and not argue with another person's opinion -- or criticism. She taught me to shake it off and to remember that sometimes haters will be haters. In 2016, I learned more about coordinating platforms and using social media to promote my writing and my colleagues' to a broader audience. I also learned how to connect with other like-minded people and share each other's work. In 2018, I learned that I could leverage the data I was creating with my online activities. I could extract, analyse and understand what was happening with my online engagement.

All sorts of data, including that generated by our social media activities, can help us find a voice and understand what we care about. It can help us recognise our values and sit comfortably in them. But it can also challenge our biases and force us to change our perspective on the viewpoints of others. With the rise of social media, our sense of place and purpose expands in drastic ways, and the data that comes from these connections can create real meaning for us. Analysing data provide much-needed insight for ourselves and our world. And when it comes to analysis, I like to include the five senses in my interpretation:

- **Hearing:** If we read what we find within our data back to ourselves, how does it sound? Is it good? Is it bad? Does it sound balanced? Is it one-sided? Does it sound complete, or do we need to find more data to prove or disprove our theories? I absolutely hate doing this as it makes me feel uncomfortable. Sometimes I read something back and I think, "Wow did I really write that? It needs another rewrite." And sometimes I write something and I think, "Wow that's good! That can't have been me that wrote it – can it?" But I understand the necessity of this step.

- **Tasting:** If it were possible to taste the data, what flavor would it give off? Good to some people and not so good to others? The results of data analysis can be more palatable to some and bitter for others, especially if the analysis disproves long-held assumptions. Data analysts must have thick skin and remember it's not their job to make all the people happy all the time. It's more important to find and present the truth than worry about sugarcoating unpopular findings.

- **Sight:** Looking at the data in different visualizations can be a fun way to see new connections and patterns. Think about how your customers, users or business leaders would interpret the data with graphics. Would they like a pictorial representation of what you discovered, or do they better understand a more simple presentation?

- **Smell:** Have you ever been doing research and get the sense that something just doesn't "smell" quite right, that something seems "off" with the analysis? Trust your instincts and take a good look at the data. Maybe the source is suspect; maybe you misinterpreted a result.

- **Touch:** How does the insight "feel" to you? Does it make you happy or sad? Excited or nervous? Disappointed or proud? Data can be used to evoke feelings in people. It can also make us look at subjects and topics from different perspectives. Do you like the data? Do you dislike the data? How clean or precise do you think the data is? What is missing that would make the data connect better with your audience or feel more "real?" These are great questions to ask yourself to produce even better insights.

- **Extra Sensory Perception**: Sometimes data or information just jumps out at you for no particular reason and you have no idea why until you dig a bit further. Sometimes it can lead down a rabbit hole, similar to the one *Alice in Wonderland*, and sometimes it can open up a whole new world, like the train that leads to the wizarding world of *Harry Potter*. Sometimes we have to let our minds go in order to understand the data and what we may find. In this instance, one person's rubbish may be another person's gold, especially since we all look at data a little differently based on our senses.

Over the years, social media has taught me that words create one source of data, a picture with words another, a video another still. There is nothing better, though, than getting a like from Hugh Jackman or Elon Musk in order to get your social media posts read by others! I have also realised that social media isn't real. We only see what others choose to show us, and we tend to share only the filtered photos and words that show our best side. We don't often see the tears, tantrums, arguments and day-to-day drama that goes on behind the scenes of everyday life. We don't often post on Instagram about being up 10 times in the night to deal with two kids under two. Or that there are three loads of washing that need doing, dishes piling up, milk spilled over the floor and no time to get it all done anytime soon. Life is messy. Life *should* be messy. But social media does not often show this side of humanity. If we neglect to document the full spectrum of life, we run the risk of denying its existence. We skew the data and, potentially, shape future generations and even Artificial Intelligence with unrealistic expectations. Nothing is ever perfect, and I kind of like life in that way, don't you? Life would be a little bit boring if we weren't thrown a curveball now and again!

While social media has changed the way we share and connect with each other and what we show the world, digital technology has also disrupted how we think of physical places, even the most important places in our lives: "home" and "work." We all know that certain places inspire certain thoughts. Mention Silicon Valley and a sense of place is already established in your head. Mention the Eiffel Tower in Paris, and another sense of place comes to life. Mention the Pilbara, the Outback, Mars and other images and sensations take root. Depending on where you come from in the world, that mental image might be slightly different. It could be based on real-life experiences or on images you've seen and stories you've heard secondhand. Our perceptions change as we learn more about the world around us *and* more about the filters that are placed on the images, articles and information available to us.

Our yearning, as humans, to belong can lead us to yearn for a physical place to make ours. After "What's your name?" we almost always follow up with "Where are you from?" "Where do you live?" when meeting a new person. But "home" does not have to be a physical place. It can be a source of comfort at work or a sense of belonging with a team. And "work" requires little more than an Internet connection these days. All of us want to have to belong, and the more we belong, the more willing we are to collaborate and learn. The challenge is in fostering that sense of belonging when the places we gather and connect move from the physical to digital worlds. It's something I've had to confront throughout my career, working for global companies from the remote city of Perth. So what are the benefits I've seen working remotely with global teams?

- Communication – You have no choice but to communicate more frequently and clearly.

- Innovation – You have no choice but to do things differently.

- Empowerment and Ownership – You have to own what you are doing as your manger may be sleeping or unavailable when you need him or her.

- Trust – You have to trust people to do a good job, as they are farther away from you and cannot be easily micromanaged.

- New Perspectives – Distance from the problem can give team members a new perspective. Distance enhances a "thinking" mentality.

- Agile and Lean Approach – Remote teams work well with this approach if they have access to digital tools that can bridge the physical distance. Think multiple cameras on white boards in individual meeting rooms to allow everyone the chance to sketch out their ideas and brainstorm.

And what are the drawbacks of working with remote teams?

- Communication – Unspoken issues will remain unnoticed and unsaid. In these instances, communication needs to be honest and up front.

- Innovation – Sometimes projects move so fast, team members have to redo work. For instance, people may develop through the night whilst others are sleeping, but when the development is reviewed it may not be what the client or the rest of the team wants.

- Empowerment and Ownership – It can be hard for the leadership team to let go of the product they envisioned or developed to allow work to be done elsewhere.

- Trust – When a team member is doing other work, it isn't always obvious to leadership. This can be very well hidden when teams are working remotely on multiple projects at the same time.

- New Perspectives – Distance and cultural differences can mean that some people are unaware that they are causing problems with a client or with the team.

- Agile and Lean Approach – Drawing and ideation can be quite hard if not undertaken in a visual form. Scheduling calls with team members all over the world can be extremely difficult as well. But there are plenty of digital tools that can help facilitate.

From an organizational and personal perspective, I think we become our happiest when we are connected. I'm "at home" when I'm responding to a thoughtful comment on my Twitter feed or sharing a laugh with a colleague over Google Hangouts. I'm at home with my remote team, with members representing places around the world, when we're discovering new insights in our data and bouncing ideas off each other to improve our product. And I'm at home in Perth, under my favourite thinking tree, in my thinking chair and every time I pick up a book with a new place to go and a new thing to learn. The key to it all is knowing when to go to your place and, as leaders, ensuring that our teams have a "home" to come to as well, where every member of the team, no matter where they live in the world, can feel comfortable to come and contribute. A place where their opinions matter and their voice is heard. A place where they belong. A place where we truly connect.

TAKE ACTION

Think about the times in your life when you feel most "at home." Where are you? Who are you with? What are you doing? That at-home feeling can be an indicator that you feel connected, content and a strong sense of belonging in that time and place. That's a very powerful state to be in, and something leaders should work to develop for their teams -- even when members' "homes" are scattered all over the physical world. Write down a few ideas of how you can recreate that sense of home for your employees.

Chapter 10

Leading Your A Team

"I love it when a plan comes together."

***-- Hannibal from* The A-Team**

When I was younger, my whole family would gather around the TV on Sunday afternoons to watch a show called *The A-Team*. In this American action series, a group of ex-special forces soldiers was on the run from the government for crimes they didn't commit. While on the lam, they take on various missions dealing with unsavoury characters around the world. "Hannibal" was the cigar-smoking master of disguise and leader of the group. "Face" was the suave, smooth-talking, good-hearted con man. "Howling Mad" Murdock was the talented helicopter pilot that swooped in and flew the team out of harm's way. And "BA Barracus" -- famously played by Mr. T. -- was the fix-it man/muscle of the group. I loved watching this team of extraordinarily talented, if quirky, individuals come together to save each other and the world. And as I began working in technology, as part of teams developing special projects and products, I started to see myself and my colleagues as something of a modern-day A Team.

Like the fictional A-Team, technology teams know what it's like to be on call at all hours of the day and night. We often have to deploy products in the middle of the night or depend on team members developing code in one part of the world, while the other side of the world (and team) is sleeping. We're always on call if something fails, and we're constantly aware that a mistake could force us to roll back to a previous version and try again. The client-facing A Team is often the key people doing the day-to-day tasks to reach the client's goals. But behind this group is a whole host of supporters, including the thinkers, the researchers and the experts who assist with their deep subject knowledge. Technology A Teams are made up of a variety of individuals with a variety of talents asked to solve seemingly unsolvable problems. But together, we make a plan and get the job done. "I love it when a plan comes together!" Hannibal always said, and I'd have to agree!

As I advanced more in my career, I moved into leadership roles, eventually becoming the "Hannibal" of my own A Teams.

Working with the best of the best can be incredibly fruitful and rewarding, but it can also be difficult to lead high-performing units. I've gained new respect for the cool-headed Hannibal after years of running my own teams. In my experience, the leader never stops. The questions are endless and the pressure intense. Personalities can be challenging to manage, and it's sometimes difficult to ensure that everyone is open and collaborative. Top performers tend to have a competitive nature, one of the personality traits that drives them to success. But this desire to "win" can be divisive if left unchecked. The leader has to keep everyone and everything balanced, all while maintaining a positive, friendly homebase for team members to work collaboratively and creatively in. If done right, the experience can be incredibly positive, with a strong sense of camaraderie developing between members and the building of friendships that can last a lifetime. But if not handled deftly, an A team can quickly become anything but.

So how can leaders of high performers keep their teams happy and successful? I've picked up a few tricks, both little and big things, that can be done to create an amazing team:

- Rotate who brings in the coffee or makes the tea or takes the notes.
- Share knowledge on a regular basis.

- Have mini challenges to encourage everyone to step out of their routine and give everyone the chance to succeed.

- Encourage sharing. A problem shared on paper is a problem halved.

- Encourage the use of white boards so everyone can see ideas and brainstorming.

- Encourage outside events and activities, like eating breakfast, lunch or dinner as a team.

- Encourage family members to attend outside-work activities too, so members can meet partners, children and friends.

- Encourage difficult conversations.

- Encourage learning.

- Encourage members to step up to new roles, even ones that might not be a natural strength.

- Teach through showing.

- Nip problems in the bud, quickly.

- Move people on quickly if they do not suit the team. Tell them sooner rather than later and then help them to find a new role that's a better fit. Bad news is best told quickly and swiftly.

- Give yourself time after a hard meeting to compose your thoughts and reflect on your feelings -- and encourage your team members to do the same. Work can be emotional, as much as we try to keep emotions in check.

- Seek help if you need it. Help could come from a counselor, a mentor, or a teacher. Never suffer in silence, and always encourage your team members to reach out for help, too.

- Have sweets (In the UK) / lollies (This is what they call them here in Australia) / candy (in America) runs, regularly. Have a jar of treats on the desk for when someone requires a little more energy to get through the day.

- Have fresh fruit on hand every day. Sometimes you fancy a piece of fruit, sometimes you don't.

- Reward staff for doing exercise before work, at lunch or after work. It's even better if they exercise with a work colleague or a family member.

- Encourage teams to "switch off" after work, to spend time with family and to not check emails or social media after a certain time of night.

- Encourage members to share books, TED talks, news articles and other thought-provoking items with each other -- and then discuss them.

- Celebrate birthdays, births, marriages and important dates from the religious and cultural origins of team members.

- Encourage the team to say thank you to each other. Have a thank you box where members can leave notes anonymously. Encourage the notes to be hand written as these mean so much more than a typed note or email.

- Encourage team members to look out for one another. If they sense something is not quite right, tell them to ask and ask again. If they still feel something is not right, ask someone else to check in, too.

- Play games together. Lego and board games can often create a fun environment to learn, especially if you have teams that love to solve problems or when language differences make it harder to converse.

- Encourage your team to work flexibly where possible -- and to leave on time. Allow them to come into the office at a time that suits their family and leave in order to get kids home from school or attend a coaching gig. Outside-of-work activities nourish and recharge us, and workers who feel support in all areas of life are happier, better workers.

- If you can, invite your team to your home once a year, so that members can see you are human, too. This will help them relate to you at a greater level and foster even stronger connections. Remember to invite everyone on the team, not a specific few.

- Communicate with members in the way they like to communicate -- text message, email, Messenger, WhatsApp. Don't forget about confidentiality issues though. Ask your team to move to a phone call if something private needs to be discussed.

- Say thank you!! A team that is appreciated and valued for its hard work is a team that will be loyal, work harder and go beyond the next time they're asked to do the impossible.

Over the years, I've gained plenty of insight into what makes a good team tick. I've also looked at ways to suss out the best-performing teams and individuals in an organisation to create the ultimate A Team for our clients. I've explored a competitive approach to finding the people who can create extraordinary things and solve challenging problems. I've blended top-performing teams together and created 24-hour challenges to see how quickly they could get things done. Other times, we've set up two or three

teams with the same project goal and watched what happened.

When you have resources to commit to these exercises, they can be

a lot of fun and yield interesting results. It's not unheard of to

come up with innovative ideas and new approaches that can lead to

entirely new products or tools. Existing products may even be

made redundant with the discovery of a technology that's better

and smarter. I have been lucky enough to play with some pretty

cool leadership ideas, including one that was recommended by the

book "How NASA Builds Teams." The 4-D System, which we

tested on one of my teams of highly skilled data scientists, helped

the team go from the top 70% to the top 90% -- and finally to the

top 99% on a global scale in our organisation. It wasn't hard to

make this change happen. In fact, it was fun! During the 12-month

process, we measured our activity and ensured we listened to each

other's needs and understood what team members were working

on. Committed to being honest with each other and making small

changes in the right direction, team members found it easy to

improve their performance and reach their goals.

While leading a team of top performers can be incredibly

rewarding, it's also important to remember that *all* of our employees bring positive qualities to the workplace. A Teams tend to attract the early adopters and innovators, which is great when a client needs something that has never been done before. But there are times when a different perspective is needed. I used to think the technology industry was no place for people who set out to meet their targets and nothing more. These people often weren't motivated by promotions, bonuses or accolades, and I found it very difficult to understand their point of view. But as I have developed in my leadership journey, I have learnt that these people are *amazing*. They know what they want, they stick to the rules and they are aware of their boundaries for work and home. They are the ones that keep a team grounded, stable and consistently delivering on the targets put in front of them. They are the ones that allow our innovators time to play. Without these people in our teams and organisations, work wouldn't get completed, quality wouldn't be kept to a high standard, and the documentation of processes could be overlooked. These steady-as-they-go employees have a variety of reasons for why they approach work this way. They may have been in the career a long time and no longer feel the need to climb the ladder. They may be happy delivering the technology they know and enjoy doing what they do extremely well. They might just feel comfortable with where they are in their life and career. These people can be amazing at teaching others, especially graduates and interns, the nuances of the workplace not often taught at colleges and universities. Team them up with an intern or graduate and you will reap the benefits. Your young employees will get a much deeper understanding of the field, and your longtime employees may just find their spark reignited when introduced to something new by somebody new to the industry.

Long ago, when I was just a little girl watching the charismatic "Hannibal" Smith lead his "A-Team," I had no idea I would have the chance to direct my own groups of sometimes quirky, always amazing individuals on totally different kinds of missions. I've learned so much from my years working with everyone from high-performing innovators to longtime career technologists. And the ultimate trick, I learned, was to give everyone a chance. Anyone can be on the A Team, something Hannibal Smith had figured out a long time ago. All they need is the chance to connect and the encouragement to reach their full potential, in their own unique way.

TAKE ACTION

Good leadership often depends on the little things -- the small acts that inspire, encourage and keep the team working toward a goal.

Think about a time you were helped by a small act of kindness or thoughtfulness in the workplace. Maybe your boss brought in donuts as a thank you for meeting a late-night deadline. Perhaps a team member noticed you were feeling down and asked to talk. What are some small acts of leadership you can add to your daily routine to keep your team encouraged?

Chapter 11

Feeling Your Way Through the Dark Times

"You gotta go there to come back."
-- Welsh Rock Band Stereophonics

In work, as in life, we are sometimes confronted with dark times, hard times, times we think we will never be able to recover from. This chapter will speak to some of my experiences with these times, and it may be triggering to those of you who are sensitive to topics such as depression and suicide. If you want to skip on to the next chapter, I understand.

I envision difficult emotional episodes unfolding like a spiral into the darkness. As unwitting passengers caught on this metaphorical slide, we find ourselves traveling down the track, farther and farther away from the light, faster and faster as we go. The one thing that can save us is the hand of a kind soul reaching in to say, "Come on, I'll help you to get out of this." Often this kind person is someone who has been down the spiral at some point in his or her life. They can recognise the path we're travelling.

As leaders, we have a special responsibility to spot team members who may be on this dark track and help them build the resilience they need to stop the descent and find their way back to the top. It is almost always darkest before dawn, an adage worth remembering and sharing at life's lowest moments. Hard times make us who we are; they help us to grow as people and realise who in our inner circle will be there for us when times get tough. They make us aware of what really matters, *who* really matters and how strong we can be in finding our way through. Hard times happen, and as leaders, we must be ready to offer a way back to the light for team members experiencing a descent into darkness.

One of my favourite bands of all time is the Welsh rock group Stereophonics. Their album, "You Gotta Go There to Come Back," always spoke to me when I was younger, even though I didn't quite realise the significance of the title. Now I do. I know it's about resilience. Resilience has been part of my nature since childhood. When I was growing up, my mum had an autoimmune disease that wasn't diagnosed until my sister and I were maybe in our early teens. Every 5 years or so after her first diagnosis, Mum was found to have a new autoimmune disease. Two memories, vivid to this day, stay with me from that time. One is calling the ambulance when I couldn't have been more than 10 or 11 years old. I remember the paramedics carrying Mum down the stairs in a wheelchair and thinking that wasn't a good sign. The other memory that jumps to mind is sitting on the side of a hospital bed with my entire family listening while the doctors explained that Mum had a disease in which the body was attacking the immune system and they were trying to find out which one. Dad used to allow us to get a chocolate bar from the vending machine when we would go to the hospital, probably to have 10 minutes alone with Mum or to speak to the doctors without little ears listening. Funny thing is, I can't eat a Dime bar now without thinking about visiting the hospital. Seeing Dad support Mum as she struggles with these illnesses has been the best example of resilience I could have ever had. I've watched his pain when he sees Mum not doing well, something that always breaks my heart and confirms how connected they are. Dad is there for Mum; Mum is there for Dad. Dad looks after Mum physically, and Mum looks after Dad mentally, and I'm sure they swap now and again as well. They are a partnership, a partnership that faces adversity together. They argue like any partnership, however they respect and value each other's opinions. They are a strong example of how a leadership team should work, and the best demonstration of resilience I can imagine.

My relationship with my partner reminds me of the connection between my parents. My partner and I have faced many challenges of our own during our time together. At 38 weeks pregnant with a dislocated and broken leg, I may have caused a lesser man to run for the hills. However my better half looked after me and, weeks later, our newborn baby, all while building three units as part of a housing investment we had taken on. He is my version of a superhero. He knows when I need to rest, and he also knows when I am being too opinionated. He's not afraid to tell me to stop, both talking and moving. He recognises that I need my distance when I am thinking or writing, and he is the first one to read all of my written blogs and books. I realise that a great partnership helps us overcome adversity and build our resilience muscles. Maybe this is why marrying your best friend is such a good idea. It gives you the tools to be resilient and allows you to have your worst moments with someone who has seen you at your worst (physically and mentally) and loves you anyway. More importantly, they will be fighting right by your side, you against the world.

While I am grateful to have the constant, steady support of

my partner, I also realise the need to reach out to professionals for guidance when a situation is too much for just the two of us to handle. Not long ago, I experienced an incident like this, one that left me emotionally devastated and took quite a bit of time to heal from. It happened when a colleague at work took his own life. I don't feel it's my place to tell his story or share those details, but I will say that I do believe, for reasons we don't understand, that some people are too good to be in this world and choose to go to another world sooner rather than later. The anxiety, insomnia, depression, differences they experience are just too much for them to handle. I respect the decisions they make in these challenging situations, and I hope their souls are at rest. But when this colleague committed suicide, I was affected by it in a big way as I had been the last one to see him leave the office. I had likely been one of the last people to ask if he was OK, one of the last chances to intervene. My emotional intelligence had kicked in when I saw him that day and alerted me to some strange behaviour. Hindsight confirmed I was right to be worried, though I could have never known at the time how worried I should have been. For months after his death, I played on repeat in my head, "I should have asked him again if he was OK." Even though I knew I had asked twice. An organisation I worked for years ago taught me that skill, to always ask three times. But that fateful day, I didn't, I only asked twice, and we agreed to catch up for coffee next week -- and I blamed myself for his death and not asking more times. I became consumed with negative thoughts on a downward spiral of self-blame. Eventually, all of the thoughts in my head turned grey, like little back clouds walking around with me. There were no rainbows, smiles or colours in my world -- not normal for the sunny girl everyone was used to seeing. I found it hard to reconcile what was happening in my mind with my typically rational side. My rational mind told me I shouldn't be so affected by this death. I did not know this colleague particularly well, though I did respect his intellect and enjoy his bubbly personality. My rational mind told me that beating myself up would do nothing to change the situation. But the rational mind isn't in control when such strong and powerful feelings are involved. Instead my brain was constantly playing the "What if?" game. What if I had said this? What if I had gone with that person for drinks? What if I had asked again if he was OK? What if I had not been there that day? The what ifs are endless and debilitating. They will consume you if you let them. I knew I needed help out of this emotional turmoil, and I turned to a psychologist to guide me. The psychologist allowed me to explore the ways I felt connected to my colleague -- we had like minds for solving problems, a love of spreadsheets, to-do lists and numbers.

We were the same age and both visionaries in the technology

industry. We had a connection, even if we weren't family or close

friends. The psychologist also helped me realise his death was a

problem I could not solve, no matter how many times I ran it

through my brain or how many times I had asked "R U Ok?" the

answer would have been the same. I had only one choice --

acceptance. I had to accept that the world had lost another good

soul -- and a super smart one at that.

When big life events happen, we tend to question

everything -- our faith, the universe, everything we know about

who we are and what we believe in. I think the questioning is

normal and part of the process of grieving and growing. What's

most important is that we learn from the event. I know I am not

superhuman (yet) and that I can't save everyone. However I also

know that my actions can and do affect others. I aim to be a leader

with heart, one people know they can come to for help. I aim to be

a leader who sticks up for team members and fights for others

when no one else will. I aim to be the leader who asks if you're

OK, and then helps you find your way out of the darkness if you

aren't. The next generation of leaders is watching today's, and they will emulate the example we set. Think of all the knowledge we have to pass on, all the tough times we've been through and lessons we've learned along the way. "You gotta go there to come back," it's true. And when, as leaders, we find our way back out of the darkness, the best thing we can do is show others how to do the same.

TAKE ACTION

As leaders, it's important for us to care about all aspects of our team members, including their emotional health. Think about difficult situations you've gone through in life and how it affected you in the workplace. Was there a colleague that reached out to help? Did a manager recognise a change in your performance and ask what he or she could do to lighten your load? Small acts of kindness can make a big difference when we're struggling. Jot down some small actions you could take to help a colleague struggling now or in the future. Keep it close for when the dark times come.

Chapter 12

Learning To Live With It

"I don't believe in guilt; I believe in living on impulse

as long as you never intentionally hurt another person.

And don't judge people in your life. I think you should live completely free."

-- Actress/Director Angelina Jolie

Guilt. It's something we all have and very few of us know what to do with. We're "weighed down," "strangled" and "suffocated" by guilt. We "carry it" with us, a moral weight that, research shows, actually makes us perceive our body as feeling heavier. No matter who you are, how successful you become or how considerate you aim to be in your everyday life, you will experience guilt at one time or another. It's a natural consequence of being human, and something that seems to be even more oppressive in today's super-fast, super-connected world. The guilt we feel now seems almost exponential. And rather than relieve it, technology fuels it. When we get a late-night email from the boss, we feel guilty if we don't respond right away. When we travel to meet a client and don't make it home in time to tuck our kids into bed, we feel guilty. If we scroll Instagram or Twitter for too long instead of doing that workout we really wanted to fit in, we feel guilt. The modern, digital world exerts more pressure on all of us and holds us all more accountable. Even our watches ping us with guilt when they let us know we're not moving enough to meet our daily step goal!

On any given day, the voice in my head sounds something like this: "I should have hung the washing out. I should have paid the bill on time. I should get a nanny. I shouldn't get a nanny. I should have messaged Mum this morning. I must attend the funeral. I must attend the birthday party. I must get a present for Sunny. I should have invited Tom and Jerry over. I should have packed the lunch boxes. I should have arranged an extra day at daycare. I should find more time to be with my partner. I should ensure I attend the next board meeting. I should book a doctor's appointment…." Sound familiar? When you become a parent, guilt moves in with your newborn and makes itself right at home: "How can I leave my child at home with my partner? How can I stay late for that presentation? How can I only do drop-off and pick-up once a week? How can I make it to the meeting as well as the presentation? How can I do the presentation *and* swimming lessons?" Those are the types of questions parents deal with day in and day out.

Leaders often take on tremendous guilt as they move up the career ladder and gain the responsibility of looking after more people in the organisation. "How can I make sure my team continues to succeed? How can I help that colleague struggling with new responsibilities? Should I stay late and work weekends to prove my commitment to the company?" Many high performers at the office try to prevent this by filing their responsibilities in this order: Make the business happy, the team happy and then their partner and their children. And forget about themselves. There's no time for that in this plan. Everyone else's needs come first. Parents often do the same thing, putting their own needs and desires aside to accommodate their children's. In my opinion, this is the wrong way around. We have to come first, we have to *put ourselves* first. If we do not resist the urge to respond to work emails immediately, if we do not allow ourselves to take a break from parenting every once in a while, if we expect ourselves to be perfect 100 percent of the time, the consequences will be severe. We will reach a breaking point, a burnout point. We will be crushed by guilt.

To put ourselves first and get back in control, we first need to break down the guilt and understand where these messages are coming from. Guilt is a learned behaviour, something we get introduced to in childhood when we're made to feel bad for misbehaving. Throughout our lives, we are taught to feel "bad" about ourselves by a variety of influences, including family/friends, religion, society, social media, even our companies. At some point in our development, the possibility of being rejected by these powerful influences makes us internalise the "rules" we feel we need to live by in order to please the judge. But as we continue to mature, we realise it's impossible to play by all of the rules. We question the source and look for ways to make our own rules, to shape life in a way that actually works for us. As leaders in our organisations, this process can be an example to those around us, especially the next generation. If they think we are invincible, that we never stop working, that we prioritise work over family life or never take a sick day when we're sick, they will try to do the same. Instead we must be transparent and show our organisations that, as leaders, we're still human. We should focus on the needs of today, tomorrow and the future by scheduling them into our diaries and making them a priority. Doctor's appointments and events that affect the health of our immediate family -- and this includes mental health -- must always come first. If we demonstrate a commitment to our family and our own health, our team will appreciate us and do the same.

As for the guilt we feel around parenting, that can be a little harder to tackle. I remember the first conference I went to as a parent, leaving my 6-month-old and 18-month-old at home for a week with my partner. The guilt was unbearable, but I found a way to focus it into something positive. If I was going to be away from my family, I thought, I better make it worthwhile. I was committed to learning as much as I could from those around me, and the experience became a really positive one when I could have spent the days beating myself up for being a bad parent instead. As my children and career have grown, I've learned to take advantage of my time away for work. I am up at 5 a.m. and in the pool. I do activities to enjoy the destination, so that when I return home, I am rested and rejuvenated. Yes, I still feel a twinge of guilt for not being there for my children 24/7, but a happy parent makes for a happy life. And for those of you who aren't parents, I know that you're aren't immune to guilt. Before I had my children, I felt great guilt every time I was asked if I was going to have kids. Those of us who cannot, will not or do not want to have children deal with this all the time, and that's the fault of society, not you. The rest of us need to learn that we should not judge or comment on someone else's life. Social media has made it all too easy to share our opinions, and it seems many of us feel almost a responsibility to "comment" on things that have nothing to do with us. As leaders, we should lead the way on a "live and let live" mentality, refusing to ask prying questions or make judgements about the way someone else chooses to live.

Guilt is a side effect of being human, but that doesn't mean it has to rule our lives. There are many ways to deal with guilt, to accept that we can't do everything and instead choose what is best for us and our families. Life is, after all, a set of choices, and sometimes the choices we make do leave us feeling heavy. But if we examine the source of those feelings and practice channeling them into something positive, we can find a way to deal with the bad feelings and keep moving ahead. As leaders, we can show those around us how they can learn to live with it, too.

TAKE ACTION

What's an area in your life that's causing you to feel guilty right now? What are the messages you've been telling yourself, and where do you think they're coming from? How could you channel that feeling into something positive? Write down a few situations and ideas for moving forward in a way that lets you live with -- and even thrive through -- the guilt.

Chapter 13

Into the Future

"The future belongs to those who believe in the beauty of their dreams."

-- Eleanor Roosevelt

What will the years 2040, 2080 and 3000 be like? What will my children see zooming by them as they catch the latest transport into the city? What will the world look like for me as I approach the age of retirement? Will I retire to another planet? Will our grandchildren be living in another galaxy? What technologies will be part of our everyday life, and what will that life look like? Will we live underground, like in the movies *Divergent* or the *Hunger Games*? Or maybe in the sky, or even the stars? Will we be able to travel to another universe or an alternate universe? Will we work? Will women finally receive equal pay for equal work? Will we consider our society a utopia or a dystopia? Imagining the future doesn't start with the question "What" but rather "What if." Perhaps that's why technologists have always had a love affair with the future. We're naturally drawn to imagining what could be. I personally love researching new technologies and hearing from experts around the world about what ideas are gaining momentum and what new and fascinating things are being developed. Mathematicians, data scientists, physicists, physicians -- they all offer insights into tomorrow, a little glimpse at how technologies we once thought were science fiction could become reality.

So what do *I* think the world will look like as I approach retirement age? Well, I hope that I will have my very own family robot to look after both myself and my partner. I hope that this robot is similar to the loveable, cloudlike Baymax in the Disney movie *Big Hero 6*. I hope it is caring and understanding and can make the process of aging more comfortable and dignified, whether in the home or in a healthcare facility. While we're being cared for by our own personal Baymax, I have a feeling we will be driven where we need to go by our autonomous cars. It's fairly clear at this point that autonomous transportation will be available in the major cities of the world sooner rather than later. We will likely go through a transition period of 10 years or so where we have a mix of transportation technologies. But some countries, maybe places like India, Japan and China, will leap ahead and go fully autonomous. With autonomous transport comes autonomous transport networks and autonomous supply chains, in which companies and countries around the world can easily and automatically move required goods and materials. While this will likely apply mainly to freight, I see a possibility of this system being used to move employees to meet worker demand as well.

While the 20th Century made us a global society, with the easy movement of people and goods around the world, the 21st may be the one to expand that model to the universe. Scientists are already considering ways to send humans to other planets with the idea that society could settle into a new home away from home. In the future, we may have the option to live on Earth or some other Earth-like planet. This scares me a bit, as I think of my then-grown children choosing to live in a different world. Would we go too? I am not so sure about that. But I now have some understanding of how my parents, and their parents before them, felt when their children made the hard choice to go to work in another country. I moved from the United Kingdom to Australia for work, for the beaches, for the quality of life and the remoteness of the environment. These same reasons may attract my children to another planet, where they will one day start a family of their own.

Just imagine having interplanetary grandchildren! Of course, in order to live on different planets, we will need incredible new technologies to support life in space. And new communication tools will have to be developed as well to support an interplanetary civilisation. Maybe WhatsApp will get an update -- AstroApp!

In my opinion, the technology of the future goes hand in hand with the commodities, the resources we will need to create this new world. The next set of sought-after commodities, things like lithium, water, air, soil (organic matter), rare plants and species, will need to be sourced, extracted, managed and optimised. And with the right technology in place, we will be able to track their properties, their location, their use and study the data to better manage precious resources. I would like to think that we will make more considerations for caring for and protecting our environment -- Earth and beyond -- in the years ahead and learn from the obvious mistakes of the recent past. Technology can play an important role in helping us be better stewards of our Earth -- and wherever humans call home next.

While commodities, transportation and the environment are important aspects of the future to consider, one thing that isn't going anywhere is money. If you have lived in different countries, worked internationally or been on holiday abroad, you know what it's like to transfer and convert money. However, what will be the currency in space? Will there be an interplanetary currency? Will we still have coins and paper money to carry around? Will debit and credit cards still exist? The best technology we have now to support the currency of the future is Bitcoin. Bitcoin uses Blockchain technology to record all transactions and encrypt the data without the need for a central banking authority. Blockchain is thought to be as big a game-changer as open-source software was last century. Will this model be the wave of the future?

Nanotechnology is another exciting development that could play a big role in the world of tomorrow. This technology makes use of microscopic materials to improve fields as diverse as construction and medicine. Some think the technology will allow us to have medicine on demand that better meets our individual health requirements. When paired with biosensors that alert us to changes and activities in our body, nanotechnology may allow injuries and illnesses to be fixed quickly and easily. At the same time, I hope humans have some control over when and how to end their own lives, or perhaps the choice to continue life indefinitely in some sort of humanoid body. Nanotechnology has a lot of exciting properties to discuss, and I'll get more into the topic in Chapter 16.

Thinking about the possibilities of nanotechnology leads my thoughts into a world where humanoids and robots exist, side by side with humans. What will life be like in such a society? What will it mean to "be human?" Where will we find our joy? In my opinion, the little pleasures of life will always be pleasurable, even if the world around us changes dramatically. Driving for the pleasure of remembering how we all used to drive with the wind blowing in our hair. Visiting local places, we once knew, perhaps through virtual reality if we're too far from home. Getting outdoors and appreciating the biodiversity of our environments. Humans always find a way to make life beautiful, and perhaps the humanoids and robots we'll work and live alongside will learn that from us. I would like to think that, in the future, the inequalities we see in society today will be nonexistent. I want to live in a world in which women and men have an equal role in society, where the colour of your skin and the nature of your job does not determine how people judge you. I know that, based on the current trends in Australia, we are far from achieving this dream. Women often undervalue themselves and what they are capable of based on societal biases. An example of this was when I didn't ask for a pay raise in a new role I was offered. My partner let me know that a "man" would ask for a lot more, so I messaged the hiring person and said exactly that. The recruiter agreed, and, because of my partner's encouragement, I started at a much higher pay rate. The experience made me wonder, how many other women in the world aren't brave enough to ask for more? Too often, women shy away from conflict, afraid to be "confrontational" or the dreaded "B word" in asking for what we are worth. We do not overestimate our abilities and we are not greedy. However, in today's world, we need to be to get our equal share. I hope the world changes by the time my daughters are in the workforce, that they can be treated fairly and compensated equally to their male (and robotic) counterparts.

No matter what the future brings, one truth remains: The future is what we make of it. It is what we can envision and what we can create when we wake from our dreams. If we believe it, we can make it happen. Technologists like me love thinking, "What if," especially when it comes to the technologies that will take us into the future and the ones will we develop to meet the needs and challenges that greet us there. It's exciting and inspiring to think about the world of tomorrow and the possibilities that exist. It's exciting to know that the work we do today is driving us all into the future.

TAKE ACTION

Do a little daydreaming about the future to exercise your brain and have a little fun. I remember a now-CTO asking me to brainstorm the applications of Blockchain on a specific industry. It was great fun to do, and it took me no more than 20 minutes to come up with some really interesting ideas. Try it yourself and see what you get. Pick a technology or an industry and push yourself to see into the future. What might it be like 20, 40, 80 years from now? More importantly, what would you *like* the future in that space to be? Set a timer for 20 minutes and see where your ideas lead.

Chapter 14

Making Friends with Technology

"A friend is someone who knows all about you and still loves you."
-- American Author Elbert Hubbard

We live in an age dominated by technology. Nearly every one of us has a smartphone in our pocket right now capable of computing with more speed and power than the PC of years ago. Many of us wear a smartwatch or activity tracker on our wrist, seamlessly connecting us to our devices and counting our steps. Our houses have bots that help us do everyday tasks, from finding a recipe to playing our favourites songs. Our weekend fun might include flying the drone over our property. Technology is everywhere and we use it all the time -- a technologist's heaven, right? Well not exactly. Even in this environment, where digital technology is ever-present and the benefits clear, it can sometimes be extremely hard for technologists to "get stuff done," to move an idea or project beyond a pilot or prototype. We know technology makes a real difference in society, for the organisations we work

for and in our everyday lives. But there are still barriers to adopting

new tools and uses. The main one? Our mindset. Many of us still

view technology with an element of suspicion, distrust. To truly

create a technologist's heaven, we need to change this thinking and

make technology a friend.

We love our friends, right? They help us when we need it,

they support us when we're going through a tough time. They

make us laugh; they make us happy -- but they also drive us crazy

every once in a while. Every friend has their quirks, their

annoyances, their flaws. But the thing about friendship is, you can

be aware -- and even annoyed -- by a person's shortcomings and

love them anyway. In fact, being aware of our friends' quirks makes

us better friends. We can make decisions about the activities we do

together -- and the time we spend apart. We can decide when to

ask for advice -- and when you'd rather not hear what your friend

has to say on a topic. Knowing our friends allows us to bring out

the best in our relationship. And technology is similar. It has its

pitfalls and annoyances. It has times when it really just doesn't meet

our needs. (We've all been there, frustrated when the laptop

decides to update two minutes before an important presentation, covering our ears when the Zoom call reverbs.) But if we become conscious of these situations, we can make better decisions as to whether the benefit of the technology outweighs the drawbacks. When we look at technology as a friend, with all the strengths and weaknesses that entails, we can improve our relationship with a tool that, let's face it, will likely play a central role in our lives long into the future.

When we choose technology for a solution in our organisations and everyday lives, we're making a judgment about how the tool can help us fix a problem based on what we know. A retail company may be in need of better forecasting capabilities, so we evaluate machine learning solutions that allow their computer programs to access and learn from data automatically. An organisation such as a police force may be having trouble accessing the right data at the right time and place, so we investigate edge computing technologies that bring data storage closer, geographically, to where it's needed. We might want to improve the user experience for our employees and design facial recognition that directs the lift to the right floor. A utility company may want to use their data to predict how customers will use utilities in the future. We choose the technology, the rules and the ways of working to meet the needs of the situation. And in doing so, we try to understand and evaluate any stumbling blocks we may encounter along the way. I've been part of hundreds of projects like these over the years and have found the most success when my clients view technology as a friend, as a tool with benefits and drawbacks that's ultimately aimed at making life easier for everyone.

With that said, I've also learned this: Just because we can do it does not necessarily mean that we should. I will pause and say that again: In technology, it is important to remember, "Just because we can does not necessarily mean that we should." It's easy, in this space, to get excited about the latest, greatest tech on the market and try to find a "problem" to solve with it or shoehorn the new technology into a real challenge we're facing. Disappointment is often the result, as clients realise the technology doesn't live up to what they hoped to achieve and the initial excitement over the investment turns to regret. Something I've found that helps guide the process is a decision log. In this written log, leaders record all of the decisions made in the course of the project, when they were made, who was present and who made the final call. The log works best when it includes discussion of potential problems, how these were assessed and what the ultimate decision was moving forward. The Cambridge Apostles, a group of intellectuals that started meeting at the University of Cambridge in the early 1800s, is famous for keeping a leather diary with handwritten notes on all decisions and discussions. This inspires me to still keep a paper log of everything I do, a cryptic set of notes for the next generation to decipher. Any great leader knows we reserve the right to change our minds (and be OK with that), revisit decisions and see if we would make the same one given what we know now. Having a log about the thought process we went through to arrive at a decision in the first place is a great starting block.

I can envision a future in which decision logs of the past become a helpful tool for projecting the future. Imagine having data at your fingertips for every decision made by yourself or your company over the last 20 years. What challenges arose? What voices were heard? What was the perceived benefit that ultimately overrode the costs? If an Artificial Intelligence had access to these records, it could understand the company perhaps even better than the humans who run it. The AI would know why decisions were made, what metrics were evaluated, and which interests won out and could automate the process, placing parameters on future decisions based on previous learning. Imagine if, with the power of Machine Learning, an AI could reason beyond the rules it had been given and understand problems even better than human leaders. An AI could recreate the decision and, better yet, improve on it. What if we could take the data from our decision logs and identify the moments that led to different outcomes? Moments of madness, genius, magic, truth, honesty, loyalty, whatever it is could be turned into a mathematical formula and, in turn, a technological advancement. (I just got goosebumps bumps thinking about the possibilities) Imagine understanding and even automating the "hunch," "a ha!" or "Eureka" moments that lead to big breakthroughs in science, in industry, in our personal lives. The entire space of decision making could be disrupted if we trust technology as a friend that has our back and share our innermost thought processes with it. Is it something that could happen in the future? Absolutely. Is it something that should? Well, that's a conversation I'm ready to log!

TAKE ACTION

Think back to a recent decision you made in your personal life or with your team. Were you happy with the outcome? Were you happy with the process? Consider keeping a decision log during your next discussion to keep track of how you arrived at the outcome.

TAKE ACTION

Grab a sheet of paper and start writing to help you conceptualise your thoughts and let your doodles flow! Try this whilst on a phone call. You'll be amazed at how your blank piece of paper fills up with words and notes. Sit back and reflect on what you captured. What insights can you gain?

Chapter 15

All About the Data

"Everything we do in the digital realm -- from surfing the web to sending an email to conducting a credit card transaction to, yes, making a phone call -- creates a data trail.

And if that trail exists, chances are someone is using it -- or will be soon enough."

-- Author Douglas Rushkoff

What makes Facebook so valuable? Is it all those pictures of your hairdresser's sister's new baby you keep seeing in your timeline? Of course not! It's the data the app collects about your age, gender, location, employer, education, friends, hobbies, political views and so on and so forth, all while you're scrolling mindlessly by those baby photos. As of 2019, Facebook generated 4 new petabytes (one thousand million million bytes) of data every day. And the company has figured out how to monetise that information for marketing, advertising and other purposes. Facebook learned early on what the rest of us have been slower to realise: Data is *the* commodity of the digital age and its value only increases with time.

Many organisations around the world have our data, and even more *want* our data. From the moment we wake up, our gadgets are collecting and storing data about us that can be used to make our lives easier, while at the same time making it easier for corporations to target us:

1. If your alarm is scheduled with Google Calendar, this data is stored with Google.

2. If you wore a Fitbit or similar activity tracker to bed, the device will log the amount of time you spent asleep and start tracking every step you take from the minute your feet hit the floor.

3. If you switch on music in the morning whilst getting ready for your day, the music app uses stored data to play the songs you like and keeps track of your likes and dislikes. The Google home hub knows you were checking out a recipe for slow-cooked curry for 6 people and likely having that for dinner.

4.   Google Maps knows to check the traffic from home to daycare, one of your frequent destinations based on data collected and stored.

5.   You top up your Smart Rider card automatically, with your banking details and travel details, including when and where you get on and off the train every day, logged.

6.   As you sit on the train checking your social media, Twitter and Linkedin, your every "like" and "click" is logged and monetised. Ads targeting your profile start displaying.

7.   When you get off the train in the city, cameras use facial recognition to collect data about you and all passengers exiting alongside you.

8.   As you enter the foyer of your building, once again your data is collected by facial recognition cameras or when you scan the key card you carry for the office.

9.   You pay for your favourite coffee via the cafe app, and the transaction gets completed via online banking. The cafe emails a receipt, along with a coupon for money off your next cup of tea or coffee.

10.   In the lift, your data is logged again as your face is recognised by cameras. If your building is using ad-

targeting technology, you'll see ads selected just for you playing on the in-lift TV. Perhaps the lift even knows which floor you're headed to.

11.   You check your emails and calendar in the lift once again, leaving behind data footprints.

12.   As you reach your desk, sensors let your company know the space is occupied and what time you arrived.

This is two hours in life as we know it today, not tomorrow or in the future, but now. Imagine how much more data will be collected from our interactions in another 5 years' time -- double at least, if not triple. Can we measure how much this data is worth? Can we measure how much someone cares about it? The value is different for every person and every organisation, but it's clear that in today's world, data can be as precious as gold.

Data impacts us and the work we do in many ways. How we think, act and deliver often depends on how we use information to solve a problem or understand a situation. We all favour a certain type of data. My own preference is locational, spatial and interconnected. Some may prefer financial, others performance information. Some may even like to see the production data. While data scientists are integral to the analysis and understanding of information, they are far from the only people using data in their work. A decision-maker in the C-Suite may look at data and value it differently to a data scientist. How we read the results is often shaped by our own perspectives of the organisation and our work, and this can be a good thing. But it can also open us up to misinterpretation in the form of bias. We can introduce bias through our data selection and analysis and, potentially, skew the results. Misinterpretation can lead to poor decision-making, which hurts our organisations. However, if we become aware of bias, we can correct it. Leaders should always be a bit wary of the data. Has it been manipulated in such a way that only one side of the hypothesis has been looked into? Has the wrong type of data been used for the analysis? Leaders have a responsibility to check, check and triple check the data that is put in front of them. Ask that awkward question about the timeframe considered. Ask what confidence value can be given to the data. Ask what your team is not showing you because they don't want to share bad news. Ensure that your teams and data scientists feel comfortable with telling you the truth, the positive as well as the negative, so you can get a balanced and truthful view of what is going on.

Data has become such an important asset that

governments and ruling bodies have become involved in legislating the collection, storage, security and use of data to protect citizens. Worries about data manipulation and security is one reason the General Data Protection Regulation (GDPR) was introduced for the EU, and why many organisations across Asia-Pacific are becoming GDPR compliant, as well. This is also the reason for Engineering Standards, including the ISO8000 for Enterprise and Master Data Management and 9000 for data quality. Without these standards, data can become disorganised, corrupted and unruly. It can become a liability. But even with these restrictions, bad actors have already left a mark. Data has been in the headlines for the wrong reasons in the last few years, with the company Cambridge Analytica becoming a poster child for its misuse. Cambridge Analytica was the UK consulting firm that acquired personal data from Facebook to create targeted political ads, most famously for the Leave EU and Donald Trump campaigns. Their techniques were wildly successful but also unethical, to say the least. Revelations around how the company interfered in national elections became a great wakeup of how much we need to value our data and how personal, identifiable data can be dangerous when used for reasons other than what's initially intended. The investigations that followed have created an undercurrent of concern and mistrust of companies around the leveraging of data.

Even though there's a lot of intrigue and headlines around the use of data in the modern world, data continues to be, well, rather boring. Processing data can be mind-numbing work, but we have to do it to derive any insights. How we process the data can also change the data, so we must "clean" the data in a way that maintains its consistency and doesn't lose granularity. When we do find that the data quality is not up to standard, it can be incredibly hard to get others to realise this, especially if those people are not data minded. The best data sets are those that have a time series associated with them, as this gives us the ability to create a virtual scenario of what was occurring at multiple places. This is when insights start to get interesting. If we have enough data points, we can use what we've learned from the past to start projecting what will happen in the future. The possibilities are exciting:

- Imagine knowing 30 minutes before it happens that someone is about to commit a crime based on patterns that replicate previous criminal activity.
- Imagine knowing the next location of a burglary to the exact house within a timeframe of + or − 30 minutes.

- Imagine being able to use the data we have around the world combined with Edge computing -- a process that aims to bring data closer to specific geographic locations -- to make decisions faster and better than ever before.

- Imagine using data to identify a new type of cancer and to pinpoint commonalities between all those infected, helping to identify the cause. Then imagine using this information to identify and notify every single person who has ever come into contact with the product or service now known to cause cancer.

- Imagine using data to understand why our brains play tricks on us, why our emotions differ based on body chemistry, such as hormones, and then understanding how to manage moods better over time.

Data has the potential to do some amazing things, and I believe that all of the scenarios I envisioned here could come to fruition within the next 5 years. Much of the data is available now. We just need to combine it with technologies like the Internet of Things (connected devices, sensors, etc.), Machine Learning and Edge computing to allow for the collection of many more data points and the ability to gain insights in faster, more strategic ways.

Another area of data that's inviting lots of discussion is the use of simulated (not real) versus actual (real) data. Simulated data has the potential to blend together existing data points with manufactured ones to create scenarios around every possible decision. It's a bit like a virtual computer game that can give you all the possible outcomes of the future. Both types of data are useful for simulations, optimizations, predictions and Artificial Intelligence. But what will be crucial is that we consider the "real" data with a bit more weight. I personally believe that it's difficult to predict what simulated data will do in certain environments. Where possible, we should consider the effects on the model that we are creating and the environment in which the data is located. My personal preference is to use real data whenever we have real data. However, when real data does not give us the extremes we need to best understand a situation, a simulated data set can supplement. Using both becomes the best option.

On a daily basis, we leverage and use data to gain powerful insights and make key business and personal decisions. To do this right, an organisation must start from a foundation of ethical values, strong principles and good data management with a strong desire to always protect the customer. Data is truly the gold of the digital age, but as future leaders, we need to treat it with more

sensitivity than we do our traditional resources. I view data as the

heart of an organisation. However, we can't use data to solve data

problems, we have to think differently, and problem solve, analyse

and assess for the outcome required, thinking about what is

broken, the people, the process, the organization, the location, the

application or the technology. Assess theses and the data problem

will be resolved. When the data is broken we think differently and

solve the problem. We must protect our data, use data and manage

data with the love and attention it deserves.

TAKE ACTION

Take 20 minutes to brainstorm some novel ways your organisation or team could use data to improve processes and make better decisions. Perhaps you could use social media data to target potential customers with a fun marketing campaign, or analyse patient data to stay on top of any emerging community health trends. There's so much potential waiting to be mined from this digital age resource.

TAKE ACTION

Speaking up and telling your truth can be scary. But there are ways to make it easier. One method is to become an expert in *your* story, using data to support your truth. Think about how your story has unfolded so far, and what data you might use to support how you tell it. Jot down a few notes now.

Chapter 16

The Little Things

"The older I get, the more I'm conscious of ways very small things can make a change in the world. Tiny little things, but the world is made up of tiny matters, isn't it?"
-- Author Sandra Cisneros

What will nanotech do to our society? How will it change the world of medicine and our environment? Could it help to save our society or destroy it? And what is nanotechnology anyway? The National Nanotechnology Initiative defines nanotechnology as the manipulation of matter with at least one dimension that falls in the size range of 1 to 100 nanometers. To give you an idea of just how small that is, 1 billion nanometers equal 1 meter, and 80,000 to 100,000 nanometers make up the width of a human hair. That's mind-blowingly small! But nanotech isn't just some weird, futuristic science. Everyday organisations, from NASA to Disney, are looking into ways to utilise this truly super tech. NASA's research has been focused around the intersection of materials and technology, on finding ways to augment materials to improve their performance in extreme conditions. As an entertainment company, Disney is, of course, focused on ways to enhance experiences, especially those delivered in virtual and augmented realities. Haptic feedback -- a vibration or rumble response in our phones or game controllers -- is nothing new, but when it's undertaken at a nano level, things get interesting. Disney has been researching haptic use in a variety of settings, from a VR Star Wars jacket that lets wearers "feel the force" to a floor system that simulates things like earthquakes caused by the Incredible Hulk smashing the ground with his superhuman fists. Nanotechnology could take these same experiences and enhance the effect by working at a level so small, it could affect our emotions, too.

The whole idea makes me think about the "magic" experiences I have encountered in my life, and how, in the future, tiny technology might prove to be the ultimate magician. Have you ever had something happen that seemed so unlikely as to be almost miraculous? Like bumping into a person you haven't seen since your primary school days on a crowded train? Or seeing your neighbour on a beach in a remote town, a colleague or old university friend at an airport as you have 30 minutes passing through before your next flight? Perhaps you picked up the phone to call someone at the exact moment they picked up the phone to call you. These moments in time leave us with a sense of wonder, of "entanglement," that some things were meant to be. What if, in the future, these chance events weren't left up to fate at all but were in fact orchestrated by big data and our vast knowledge about individuals' movements and routines. What if nanotechnology is developed to bring more of us together, to nudge us in different directions and change our behaviours ever so slightly to reach more serendipitous ends? We might never know if the moment was brought to us by chance or technology, making the magic even more special.

There's a term for the type of situation I'm describing, Quantum Entanglement. In physics, it's the idea that entangled particles stay connected even when separated by great distances. Occasionally, "lightning strikes" and rare events happen due to the colliding of ideas or events, warping time to create a moment as sharp and exact as a lightning bolt. What seems to us a miracle is in fact caused by natural particles that connect us to each other and to other things. My understanding of these mind-bending concepts comes from the Australian television programme *Catalyst*, which did a series, "How to Build a Time Machine," and the book *The Improbability Theory* by Sir David Hand. In the television show, the hosts investigate the strangest things the quantum world has to show us. In one example, a physicist uses two dice linked together to simulate something like sisters separated by geography yet still in tune with each other's feelings. Yes, we have all heard of this concept -- something akin to Twin Telepathy -- and yet it has never fully been researched. My sister and I do this regularly. One of us will feel something and the other will show the same emotion, no matter how far apart we are in time or place. This, I also believe, can happen with partners, parents and children, perhaps even with animals. How do we know that our dog would like a treat, by his behaviour, or is it something else?

Initially this concept was termed by Albert Einstein as "Spooky Action at a Distance." Evidence showed that entangled photons may be able to choose or predict their state, an idea so different to the norm that even Einstein and another early architect of Quantum Physics, Erwin Schrödinger, ultimately found it too outlandish to be the answer. But Entanglement is an important aspect of reality that, with better understanding, we may be able to use to improve our interactions and communication -- and there's some exciting research on the topic. In a study released just this year, physicists from Austria and the U.S. observed quantum entanglement among "billions and billions" of flowing electrons in a quantum critical material. The researchers noted the new technologies in computing and communication their findings could unlock. Imagine the use of Entanglement to provide instant communication. We'd no longer need fibre optics, 5G and wireless technology. We could just use Quantum Mechanics to transmit information instantaneously. Take this one step further and imagine if we could apply the principles found in Quantum Mechanics, specifically Entanglement, to Artificial Intelligence. We could create a replication effect and control AI, giving an emotion or state to the AI. We could expand this control to robotics, which would be extremely powerful in changing the way we handle space exploration and, perhaps in the future, interplanetary travel.

The concepts of nanotechnology, Entanglement and Quantum Mechanics can be incredibly hard to wrap the mind around. There's a reason Einstein was considered a genius! But thinking about scenarios like this is a way for leaders to ensure their organisations stay ahead of the curve. You can bet the innovative companies of the future are thinking about nanotechnology and other developments at a whole new level. True leaders need to take risks that will allow them to adopt technologies faster than the competition but taking a chance doesn't necessarily mean fully understanding all of the exciting developments taking place. New maths, new natural laws, new physics concepts and technologies are being discovered all the time. It's crucial that leaders keep up to date with as many changes that they can, but even more important for them to be *open* to change. We cannot assume that what we learned in school 20 years ago is still true today. The theories may have been debunked or reinvented, the science improved. This fact is going to be truer tomorrow than it is today. One way to stay on top of new ideas is to foster relationships with universities, with bright and developing minds. Leaders and organisations should ensure they have direct connections with universities, encouraging new research, understanding the latest developments and hopefully bringing the brightest minds into their work. The next Einstein is out there with ideas sure to rock the world. As Einstein himself said, "Logic will get you from A to B. Imagination will take you everywhere." Add in the developing technology of nanotech and beyond, the "everywhere" Einstein envisioned becomes even more limitless.

TAKE ACTION

You don't have to be an Einstein to imagine how technologies like nanotechnology could change and even disrupt entire industries. Put your "genius cap" on and think about how this tiny tech could have a big impact on your organisation now and in the future.

Chapter 17

Reimagining Reality

"Reality leaves a lot to the imagination."
British Singer -- John Lennon

As a kid, I was fascinated by and obsessed with computer games. The flashing lights, the points racking up, the need to try and try again to figure out how to complete a level. I loved it. And I can see this enjoyment in my youngest child, how her face lights up when she sees an old-style pinball machine. When I was a kid, "screen time" wasn't a term anyone used, but as a 21st century parent, it's something I think about quite a bit. How much time should I allow my children to play on their devices and explore worlds -- both real and imagined -- online? How can I, a technology professional, model the right balance of time spent looking at screens with time spent in the real world? As technologies like augmented reality and virtual reality combine with haptic sensing and more immersive experiences, the possibility to

147

fully escape into a virtual world seems closer than ever. Is this a good thing, or are we headed toward the dystopian future movies like the *Matrix* warned us about?

We undoubtedly benefit from engaging with virtual realities. Visiting a virtual world allows us to understand how different people live, creating empathy for and engagement with those who would otherwise be perfect strangers. A few years ago, the BBC built a much-lauded virtual reality story around the Syrian refugee crisis in Europe. The piece told a hugely important political story from the perspective of those most affected in a way that brought empathy to their plight. We often talk about "walking a mile in someone's shoes," but "seeing" the world through the eyes of another can be even more life changing. Imagine experiencing how an act of bullying or another trauma made someone feel. What amazing insights could be gleaned by therapists, policy makers, family members and others trying to help an individual or solve a societal problem. Imagine, as the leader of your business, popping on an Oculus Rift to see the company from an employee's point of view. You could better understand how employees feel about their work, the company culture and their view of your own leadership. What a dramatic change this would be from the static, boring survey questions human resources ask of employees today.

Business leaders of the future may use virtual reality to create

models of entire projects, such as a networked transport system or

the next city tower. Not only would the technology let you see how

the project would unfold, but you could "feel" the wind in your

hair, the way the sun filters down through the landscaping. You

could experience the end result before a single shovel broke

ground.

As we head into a world post-Covid 19, virtual reality has

the potential to fill the gaping hole left by the virus in the travel

industry. Instead of flying to Venice, perhaps we book a virtual trip

to explore city destinations. We could "visit" the Pyramids in Cairo,

the Louvre in Paris or the polar ice caps before they melt, without

leaving home. At the moment, no video or even augmented reality

can capture what it's like to, for instance, walk on the moors of

Dartmoor. We were watching a documentary the other day about

it, and my partner was shocked at how quickly the fog and rain

came in, engulfing the camera man. I reminded him of an

experience we'd had walking on the Welsh Moors to visit King

Arthur's Stone and he suddenly understood how cold, wet, chilling

and scary the experience would be. He understood because he had lived it before. But if he hadn't had that real-life experience, a virtual one, enhanced with sound and sensing technology, could have been just as helpful in setting the scene. VR should also be seen as an important tool for education, especially in the post-Covid 19 world. Older students could experience the history of the world wars; younger children could "travel" to see the geography of unique places or even visit with now-extinct animals. And if another virus forces schools to close for a period of time, Zoom or Teams classes could be replaced by VR, bringing a student's teacher and classmates right into the home.

As we further develop and engage with virtual reality, we must always stay aware of the primacy of reality; nothing can or should replace the real world. In order for this to remain true, we will need new regulations and laws around how we use virtual worlds, how we maintain and protect people's data and how we deal with ownership over virtual places. In the distant future, I can imagine people creating entire lives within a virtual setting. The rules that govern our physical world will need to be translated for this new frontier. How we interact with each other and the places our VR creates, how much time we spend in this strange space between real and not -- all of these guidelines will need to be addressed. And for someone like me, someone naturally drawn to new and exciting technologies like the video games of my youth, society must consider the risk of people disappearing altogether into this virtual reality. In today's world, people seek help for video game addictions, addictions to social media and their smartphones. How can we prevent the same problem in VR? How long would we be allowed to stay in the VR world? How much is too much? The question of screen time just got a lot more complicated.

While virtual reality is a way to bring the natural world "to life" through technology, another fascinating and promising development looks at ways to enhance technology through natural laws. It's called quantum computing, and it's based on the very complex ideas of quantum physics -- Einstein-level science that leaves most of us feeling befuddled. You may have heard the term quantum computing before; the idea has been researched for decades. The goal is to use quantum-mechanical principles to evolve computing to the next level. This has the potential to increase processing speeds by 1 billion times and greatly enhance problem-solving capabilities. No longer would we be waiting on computers to process. The computers will be waiting on us to tell them what to do. This is certainly a paradigm shift. Computers are not used to waiting for us humans, although it's not like they have the virtue of patience anyway.

Discussions about quantum computing are not new but what is exciting now is how close we are to making it happen. Last year, Google claimed to have realized "quantum supremacy," though some critics said the company hasn't fully tapped into the power of modern supercomputers. Pawsey Super Computer Centre here in Perth, WA is pretty awesome and fascinating to see the processing and case studies that the super computer is being used for especially now. IBM has a quantum computer that can be accessed online. Anyone can log on to play a card game against the supercomputer. The technology is said to be incredibly powerful but not yet reliable, and researchers agree there's much more work to do. Quantum computing is something of a contentious, competitive field, but, in my view, it shouldn't be. Researchers should work together and support each other since the technology has the potential to make a tremendous difference to our future. We know we face incredible challenges posed by climate change, viral pandemics, political instability and more. The problems are just too big and complex for us humans to wrap our heads around. And while today's computers can help us model and understand the issues, quantum computing could be a way to help us *solve*

them. This technology has the power to look at problems of a large

magnitude and evaluate an infinite number of solutions, quickly

and efficiently. Quantum computing could guide us in enacting

data-based strategies to save lives and change our future for the

better. Now more than ever, I think, we *need* this technology. We

need a way to find solutions to the incredibly complex problems

we're facing -- and reimagine the reality of our future altogether.

TAKE ACTION

While virtual reality and quantum computing seem a bit too sci-fi
for most of us to wrap our heads around, they do have great
potential to change the way we lead our businesses -- and our
teams -- in the future. Take 20 minutes now to consider a problem
you're facing in your business or workplace. How could you use
VR to help solve it? What question would you ask a quantum
supercomputer to answer for you? Though you might not have
access to these technologies yet, thinking about your problem from
this perspective may just lead you to a creative new insight.

Chapter 18

Breaking the Rules

"To study music, we must learn the rules.

To create music, we must break them."

-- French composer Nadia Boulanger

At the end of my workday, when my two children and I are heading home from their after-school care, I always ask the same question in an upbeat, excited voice: "Who went to the naughty corner today?" Both of my girls excitedly think about it and put their hands up if they spent time in time out. We talk about the scenario that got them in trouble and see if their actions were deserving of the naughty corner or not. My intent is to teach them that rules are just that -- *rules*. Sometimes, if the intention is good and the consequences manageable, it is ok to break them.

Sometimes, in fact, you have to break the rules to progress in life. I am trying to get my girls to learn how to do a cost-benefit analysis of their behaviours. I want them to learn how to weigh the pros and cons of following the rules. I also think it's important for them to process some of the not-so-pleasant things that happened during their day. As psychologists well know, telling our stories can help us understand our emotions and process events, and I think this is especially helpful for children.

Once, my 5-year-old, whilst writing a letter to the tooth fairy, said "Mummy I did go to the Traffic Lights (what they call the naughty corner at school) today!" She smiled at me with a huge grin and I said, "Oh, ok what for?" The answer was that she wanted to share a tiny piece of chocolate with her friend. Unfortunately, the teacher spotted them, and when the teacher told her they were not to share lunches (due to allergies), she and the friend had to go to the corner. When I asked "Did you go together?," she said they sat at opposite ends. When I asked, "Was it worth it?" She said "Yes!" and her face beamed. I could tell this minor transgression created a bond between my daughter and her friend, and the massive grin she gave me when I asked, "Did you enjoy your chocolate?" was beyond doubt the best response I could get. My daughter's teacher taught her the important lesson that rules are enforced, and they often have a valuable reason to exist. But our discussion showed her the power of knowing that rules should sometimes be bent or even broken. Do I want to raise a child who is kind? Absolutely. Do I want a child who is going to follow the rules? No, not really. I want to raise a child who uses the rules as a guide -- and her brain to decide the right thing to do.

This philosophy applies to the intersection of business and technology, especially when dealing with new technologies that don't quite meet the existing structure of an organisation. Often when new technologies are deployed, the old rules around data, commercialisation, security, even deployment are challenged and must be changed to facilitate adoption. For many organisations this can be a scary prospect, especially if the company is a risk-averse one. Tactics such as "sand pitting" and "ring fencing" can be used to manage risk and keep the new technology separate from other systems until it has proven itself. Using dummy data rather than live data is another way to prove a concept works, as is starting small, with a prototype targeted to a department or employees who better meet the profile of early adopters. For instance, your marketing department may be more open to a new technology than your accounting. Sometimes it's best to allow the idea or prototype to develop naturally, at its own pace, and just monitor the progress. Or you may want to put together a product roadmap that details the vision, direction and steps for deployment. This can be a great way to chart the course, identify potential areas of concern and realise the processes or technologies that will become obsolete with rollout. When you're going through this process, you're also likely to identify the "rules" that threaten to slow or stop progress altogether. How you deal with these roadblocks can mean the difference between success and failure.

In my career, I've found that the fewer rules we break or change, the easier it is to implement a technology. So, we must choose our rebel moments wisely. Some technologies are easier to get your head around. For example Google Translate, Google Assistant, smartphones and smart watches fit into our existing lives so easily that we hardly notice them anymore. My daughter said "hello" to Google the other day before me, asking Google for a song as soon as she walked through the door. However, if you miss a few generations of technology, it's harder to play catch up. A friend of ours, who is 80, asked us to help her do online banking from her phone. After obtaining the latest smartphone, (her last phone was a basic Nokia), the first challenge we had was teaching

her to unlock the screen and explaining how apps work. We then

had to explain the difference between WiFi and data and give her a

crash course in the Internet. We contacted the bank to set up her

online account and came up with a way to help her remember the

password. After all of that, she managed to access online banking

for the first time. The challenges seemed endless because so many

things had changed in the mobile phone world since she had

purchased her last phone. But to her credit, she stuck with it. She

could have given up and decided to just keep going to the brick-

and-mortar bank the rest of her days, but she opened herself up to

change and got there in the end.

Change is scary, and too much change can be

overwhelming. I think we've all felt that in recent months as our

world was turned upside down by Covid 19. In a matter of weeks,

we were confronted with change on an enormous scale, and we

didn't have the choice to adopt or not. Our health and, potentially,

our survival depended on us accepting these new rules of society.

In uncertain times and situations, rules can comfort. We can't

control the spread and tragedy of the virus, for example, but we

can follow the rules around socialising and washing our hands. We know children respond best to routine, to knowing what comes next. And I think adults do too. We like to know what's down the road, what to expect. And one way to ensure this is to follow the rules you set for yourself or that you adopted from your parents, your religion or your culture. If you stop and think about your day-to-day routine, I'm certain you'd be surprised at just how many implicit and explicit rules you abide by, from eating a healthy cereal to please your heart doctor to traveling down the road at the speed limit to please the police.

Rules have meaning and reason to exist, but it's in the moments when you bend the rules or break them altogether that life gets exciting, that new possibilities arise. One of my colleagues has recently done just this. She tossed aside all the rules of her existing life to set a new path forward. She left her job and paid off her mortgage and is now months into her own start-up design consultancy. I admire the courage she had to give it a go, to take a stab at living life on new terms. That courage is what I hope I'm reinforcing in my daughters when I ask them to tell me about the times they break the rules. It's that courage I try to inspire as I help companies confront and accept change, sometimes breaking rules in the process. And it's that ability to be brave even in the face of fear that I call on when my own life demands a rebel moment.

TAKE ACTION

Take a few minutes now to do a cost-benefit analysis like the one I encourage my girls to do every day. Did you break any rules today? Did you want to? What rules in your personal or professional life are holding you back? Consider both explicit ones, such as company policy, and implicit ones, such as company culture. If you could break that rule, what would you gain? What would your company gain? Is it worth a rebel moment?

TAKE ACTION

While virtual reality and quantum computing seem a bit too sci-fi for most of us to wrap our heads around, they do have great potential to change the way we lead our businesses -- and our teams -- in the future. Take 20 minutes now to consider a problem you're facing in your business or workplace. How could you use VR to help solve it? What question would you ask a quantum supercomputer to answer for you? Though you might not have access to these technologies yet, thinking about your problem from this perspective may just lead you to a creative new insight.

Chapter 19

Leading the Future

"Be strong. Be kind. We will be OK."

-- Jacinda Ardern, New Zealand Prime Minister

My family is filled with builders and fixers. Whether it's building houses, engines, communication systems, relationships or networks, my family has done it all. I grew up in the UK's equivalent to Silicon Valley, the M4 corridor where all of the country's big technology firms were based. Everyone we knew had a connection to the tech industry. My father was trained as an apprentice in one of the world's first data centres for the oil and gas industry. He worked his way up to become network architect and solution consultant. He was also a farmhand in those days, never fully committing to the white-collar lifestyle. Growing up with Dad "on call" was an opportunity for me. I'd go with him to the office to solve a problem or help him type the relevant code into the computer to ping the network address. These moments gave me great insight into how important technology was to business -- and how much manual effort it took to keep it running in the early 1980s. Because Dad worked in this space, we were lucky enough to have access to the latest gadgets at home. Including Commodore 64, also every latest game that came out, but most of all was the energy and the excitement when he had come from the latest conference and hearing someone speak about the future. I'm sure my interest in technology took root in these experiences. And because my parents raised their two daughters to believe they could do anything and be anyone, I never doubted I would succeed in this career, even if it wasn't easy.

The face of leadership in technology and many other industries at that time was very much male and often white. But that didn't deter me from pursuing my passion. If anything, I took it as a challenge. We were brought up to treat everyone equally and with care and above all to treat others the way you wish to be treated with care and kindness. I purposely sought out women in leadership roles, fervently believing, "if you can see them, you can be them." I looked for female mentors to learn from, to watch and emulate -- and I serve as a mentor today to the next generation, passing on those lessons. Leadership has changed quite a bit since I started my career, and the value of women is being more and more recognised. We've had powerful women run tech companies, from Meg Whitman, formerly of HP, to Facebook's Sheryl Sandberg. We've seen countries thrive under the leadership of female prime ministers, such as New Zealand's Jacinda Ardern. Ms. Ardern helped her island nation navigate the trauma of a horrific mass shooting and successfully stamp out Covid 19 over the course of one very eventful year. She did all this while being a new mother, too. The world has celebrated Ardern's approach to leadership, one based on kindness, community, compassion and strength. She speaks openly about kindness being a strength, sadly a novel idea for the male-dominated, capitalist world we live in. She proves that traits traditionally thought of as "womanly" -- collaboration over competition, compassion over indifference -- are not weaknesses but values that female leaders bring to their organisations. I see Ms. Ardern as a model for the leader of the future, and I, too, walk that path.

"Emotional intelligence" is the term many apply to this style of leadership. It's the idea that knowing your own feelings and being able to tap into and understand the emotions of your team members, customers and clients makes you a better leader. These are people who show empathy, listen when a colleague is having a tough day. These are people who know motivation works better than discipline, and that when you're called on a misstep, a heartfelt apology is required. These are people who treat people as people, who see others -- and themselves -- as fully human beings with strengths and weaknesses, good days and bad. It's somewhat unbelievable to me that in the 21st century, we're still learning to embrace people as people. But that's where we are and, fortunately, we're headed in the right direction. My natural tendency is to think positively about the future, even though we've experienced some scary and very sad events in recent months. From the coronavirus pandemic to the death of George Floyd at the hands of police in the U.S., events like these happening around the world have been discouraging and heart breaking. Yet at the same time, we've seen the best in people. We've seen volunteers bringing groceries to our most vulnerable populations to protect them from getting the virus. We've seen people around the world standing together in protest of racism and the institutions that propagate it. I've always seen technology as a tool for creating the type of future we want. It gives us the power to act creatively, to do things differently and truly change our world. It gives us the tools to understand and solve problems we have created in the past. The leaders of the future have the potential to embrace technology, as well as the softer skills of emotional intelligence, to bring about a world that treats us all a little better.

Of course, there's always a flip side, and as good as technology can be, there's always the possibility that it's used with bad intentions. The scary use of "deep fake" media to prevent us from knowing what's real and not in the news. The overreaching use of surveillance tools to control populations, identify and prosecute people for protesting against the government. The insidious use of social media to manipulate people, radicalising their political beliefs and even changing the outcomes of elections. These events aren't just some plot in a sci-fi novel; they've really happened in recent years. At the moment, there is no global force to stop them, and some political leaders have even embraced them. How are we going to ensure our society continues to uphold the truths of humanity and our right to a fair and just world? That's something we all need to grapple with going forward.

With any technology, with any situation, there is always a choice, a moment when we decide to work for the benefit of others or to the detriment of others. If we decide now, collectively, to hold ourselves accountable, to judge our actions, how we use technology and its consequences, we may have a chance at a hopeful future. This is even more important as we develop more sophisticated Artificial Intelligences and robots that become embedded in our society, almost human-like. Our laws need to be updated at a local, global, international and, eventually, interplanetary level to ensure that everyone is protected from the nefarious use of technology, now and going forward. A wise leader once said to me that the world of the future will be divided by the "haves and have nots" in reference to higher education. In my opinion, the world of the future will be one in which technological experience, training and understanding becomes the dividing line. Those who know how to code computer programs, develop AI and analyse data have more power to create the world we'll live in tomorrow. But all of us should be involved in the process of forging that future. Now's the time to think about the guardrails we need to put in place, the measures we need to enact

to protect ourselves going forward. And now's the time to

embrace the value of female leaders like Jacinda Ardern, "Be

strong. Be kind. We will be OK."

TAKE ACTION

When you take on a project that will eventually change how your team or organisation currently does something, stop to think: How will this affect our people? Talk to those involved, get their viewpoints, understand their hesitations. By taking people along with you on the journey to change, you'll have more buy-in and support for the process.

TAKE ACTION

Think about how you can help kids in your area learn technology, so they do not become the "have nots" of tomorrow. Are there organisations you could volunteer for, school programs you could assist? Maybe you and your company could start a mentorship program. Make a list of ways you can make a difference and lead the future.

Chapter 20

Saving Our Earth

"I used to want to save the world. To end war and bring peace to mankind;

but then I glimpsed the darkness that lives within their light.

I learned that inside every one of them there will always be both.

The choice each must make for themselves -- something no hero will ever defeat."

Fictional character -- Diana Prince, Wonder Woman film

I've noticed a shift in the workplace in recent years as more "youngsters" join our ranks. The Gen Yers (also known as Millennials) and Gen Zers come into the workforce with so much confidence in their abilities, with leadership traits already starting to form. Many of them have strong, passionate views about social and political issues. They care about equality and equity in society, in politics, in the economy and in work. They advocate for social justice and civil rights. And they yearn to protect the environment, with a zest and energy that's changing business on a global level. We can learn so much from these generations that have been change-makers since the get go, just as they can learn from those of us who have been around a little longer.

I'm not sure when the discussion around climate change shifted from a mostly academic exercise to one everyday people participate in and care deeply about. But the inflection point definitely occurred during my time in the workforce. When I first started working in technology, environmental concerns were present but not big enough to guide or shift business decisions. Having studied Geography at University the environment was a consideration for every module. However, on my entry into the workforce it was only evident in pure environmental projects and then it was about fixing the outcome not preventing the problems caused in the first place. A company might talk about recycling or use a recycled material. But that would likely be the extent of their environmental awareness. That's no longer the case. In recent years, consumers have successfully called on companies to change their environmental practices for the good of all of us. The power has shifted. Eco-conscious consumers, many of them Millennials and younger, are looking for more than cheap prices and high-quality goods. They're looking for companies and brands that align with their political views and social values. Things like "fast fashion," in which companies mass produce trendy clothing at cheap prices with poor labor and environmental practices, have been called out. In its place, consumers express -- through their hard-earned cash -- a desire for sustainable manufacturing that promotes the responsible use of materials and labour.

The tech industry has a big burden to bear in this new frontier, as the gadgets we love consume electricity (which, in most countries, still comes primarily from burning coal). Cellphones and laptops depend on batteries and parts filled with materials mined from the earth. Many of these raw materials can be found in my home state of Western Australia. Our rich natural resources include lithium, nickel, cobalt, aluminum, vanadium, zinc, manganese and graphite -- all the elements needed to make lithium-ion batteries. But we've also got an abundance of wind and sun here. This combination makes us well-placed to transition to cleaner forms of our energy, and our leaders see the potential to help the environment while creating local jobs. "Here in WA [we] are so well placed, we're positioned like no other jurisdiction in the world to make this transition," says Chief Scientist of Western Australia

Peter Klinken. While local governments look to embrace the future of energy, companies are searching for alternative ways to produce the technology we need, from solar-powered laptops to phones made of cornstarch bioplastic. Invention is indeed the offspring of necessity, and smart leaders know their organisations must be friendlier to the environment in order to attract customers of the next generation. These customers don't mess around. If a company fails to adapt, it's going to be "put on blast."

Companies have a big role to play in saving our earth, but I also foresee a future in which every single one of us is aware of our personal carbon footprint and asked to make better choices based on that data. I envision us wearing a Fitbit-like device that tracks our travel, our consumption habits, where our food is coming from and how it's been produced and so on. The device then crunches the numbers and gives us minute-by-minute feedback on how our activity affects the environment. Perhaps we each get a monthly stipend of "carbon credits," and if we travel out of town, take our personal vehicle instead of the bus or eat a meal of lobster imported from Maine, USA, we get docked more credits than if we made sound environmental decisions. By the end of the month, if we have no carbon credits left, we're stuck waiting for the next allotment. We stay home, cook food for ourselves and do a puzzle by candlelight. (This sounds like bliss to me) When our credits renew, we can hit the town again -- or save them up for a big trip a few months down the road. I love this idea because it combines several of my favourite things -- technology, data and rational decision-making. If we can visually see the impact we have on the environment -- not just in some hypothetical way, but with real numbers and real-time data -- we can change our behaviours at a micro level which will have an impact together on the macro level.

The recent lockdowns due to the Covd 19 pandemic showed us all, very clearly, how our everyday movement affects our Earth. When we stopped commuting to work, stopped eating out at restaurants, stopped flying across the world at the drop of a hat, the earth had a chance to heal. I'm sure you saw news footage of the clear waters of Venice canals, the blue skies in Beijing. These stories were perhaps the only good news to come out of this chaotic time. And what they showed me is that humans have the power to change course, to stop climate change, to heal our world. This problem we face is not insurmountable. We don't have to feel helpless, we just need to act. The tools we use to save our earth might be as simple as staying home one extra weekend a month. I'm really excited to see what sort of research and creative ideas arise from this pandemic, and I'm certain technology has a big role to play. I'm even more sure that the leaders of tomorrow -- all of you Millennials and Gen Zers who have already made such an impact on our world -- will be the ones to solve this problem, to finally confront climate change in a way that doesn't destroy our economy, but rather fortifies it for the future. And I'm so excited to be part of the solution.

So there's a choice, a conscious choice each one of us can make for ourselves, small decisions that everyone can make as we go out into our world, do we buy or do we grow? do we walk or do we drive? do we throw away or do we recycle? Changing our decisions from conscious so that they become unconscious in our family and become part of our every daily life are important for our future. Simple decisions that can change the world. Are you ready to change the world?

TAKE ACTION

To save our earth, we all need to change our behaviours. Take stock of recent decisions in your organisation or personal life, from purchases you made to travel you planned for yourself or your team. What could you have done differently to make those decisions more environmentally friendly? What could you do going forward to build that question into your decision-making process?

Chapter 21

The Pandemic

"Look at the stars. Look how they shine for you."

-- Singer Chris Martin, Coldplay

What will our children learn from the Pandemic of 2020? What have I learned from the events that occurred and unleashed a wave of change in the way we live and work? Hundreds of thousands of deaths experienced worldwide in just a few months, and the number isn't slowing in some hard-hit places. Here in Western Australia, in Perth, what's often called the remotest city in the world, we did well to lockdown movement and quickly reduce deaths and numbers of new cases. New Zealand has done the same, and now there's talk of a bubble being established between our countries as a way to encourage travel again. Our new "normal" is so different from the world we lived in just months ago that it's hard to believe what I am writing! And yet we are the lucky ones. We can look at the stars and they can shine for us. Some people and countries in the world are not so lucky and find themselves deep in the throes of this terrible pandemic or mourning unimaginable losses.

Months before the pandemic hit, my family had a big life change that actually made us a bit better prepared for what was to come. In late 2019, we moved to an off-grid home outside the city of Perth, at least a kilometer walk from our nearest neighbor. The second property we own, located in the city, had been loaned for free to family as they needed our support. Before the pandemic, people thought we were insane to make a move to the "middle of nowhere". "Are you preparing for the end of the world or World War 3?, they would ask. After the pandemic, they were in awe of what we had achieved, inquisitive. Our remoteness easily allowed for social distancing and also shifted our mindsets around how we shop and travel and perform daily routines. When you live far away from the store, you make the most of every trip and really stock up on your supplies. When the pandemic hit, we were already used to buying the necessities in one "trip to town."

The week of March 16 was when everything changed for us. The government told us to work from home when possible, and our children were to be kept at home. We heard of three friends in the United Kingdom pass away in a 6-week period,

though Covid 19 was never mentioned. We don't know if they
were even tested. Here in Western Australia, borders -- both local
and state -- were shut down. We were screened if we sneezed or
coughed; flights were cancelled. We also dealt with an incident on a
neighbor's property, where my husband had to give CPR and I had
to drive to meet the ambulance with one injured little girl. (In a life
and death emergency situation, Covid-19 rules we're not adhered).
This certainly scared all of us as a family. Talk about adding stress
to an already stressful time! During this period, we moved into our
off-grid home from an old 1970's caravan we had been living in
onsite. Socially isolating while off-grid is an adventure. Our two
solar panels had to work hard to power the fridge, three laptops
and our mobile phones. It was necessary to conserve energy
wherever possible to keep our "workstations" powered, and we did
it! Our battery room is now our favorite room in the house, and so
far, we haven't dropped from 100% capacity in the battery storage
bank. (We did test it to the max and found that putting the kettle
and heater on at the same time will kill the power. Good to know!)
I have "shown" my work colleagues around the house on video
calls, and they've been fascinated with the idea of going off-grid.
My children are also amazed that Mummy's laptop and their own
devices run from the Sun.

Working from home, I found myself walking our property whilst talking on Zoom, Teams, Skype, Messenger or Chime calls. Choose any platform, I've mastered them all! Talking while walking helps me think logically and strategically. The trees and pathways on our land provided a welcome release from the stress of challenging work calls and the changes currently happening in our world. It didn't seem possible, but I really do appreciate technology more now. I appreciate the telemedicine, the virtual conferences, the teams around the world that support essential services and payment transactions, the clouds and the creativity that the technological world can bring us. The kids seemed to be glued to their laptops in-between the "activities" they had been given by school, but apps like ABC, Reading Eggs, Seesaw and so many more kept them engaged and learning. They are now able to navigate and understand new user interfaces quickly and intuitively. At the age of 5 and 6, they can organise a call with family members on the other side of the world, record a video for their teacher and take photos to describe what is happening in their lives. Children adapt quickly, and mine have stopped asking if they can go on the play equipment at the parks. They instead ask to go to the beach or for a walk or to the tree in our bottom paddock where we hung a swing. They have been taught not to hug and not to shake hands.

They now know how to do a virtual hug. While it's amazing to see their resilience and ability to adapt, I do wonder what effect the pandemic will have long term. I hope it makes them stronger, more capable of using technology -- but also more grateful for human interaction. I hope it makes them more caring, kind and creative. And I hope few children are left behind because they can't access the technology today's learning requires.

The lives and dreams that were taken or disrupted in this horrible time will haunt me for a long time. Leaders have had to make decisions that broke their hearts in two but ensured the survival of their business, their people. Workers have voluntarily raised their hands to help, offering to go part-time or volunteer for redundancy. People have acted heroically and humbly when the situation called for it. You appreciate those actions more in times like these. You appreciate the teachers, the patient ones that get that not all activities can be completed online and that some children would be happy to sit and draw or paint or play or bake instead of looking at a screen. Children are all different, and for the mums and dads out there who have had to work, homeschool and deal with loved ones being sick and even dying, life has been difficult. I appreciate the teachers who got this. I also appreciate the emergency services workers, the paramedics, nurses and doctors who cared for us all during this time. And those people who continued serving us, even when it was risky, on the new frontlines -- our grocery store clerks, the fridge fixers and postal workers, our rubbish collectors. Without them, life would have ground to a halt. I appreciate the farmers and the growers who kept us fed. I also appreciate the freedom we have here at our farm, the space we have to roam.

Because Western Australia locked down so quickly, we were able to return to something closer to normal life more quickly too. I've started spending time in the office again, but it's taking time to feel comfortable with regular life. Some people are itching to escape to nature, the place we have been "stuck" these last weeks. The government now allows us to gather in groups of 10 and camping has been allowed. We have been inundated with people wanting to come and connect with nature and camp on our property. Never has my husband responded to so many requests! We don't charge much, not even enough to cover our own costs. But we're happy to provide the service. And in our new off-grid life, we're able to live more simply financially. We have a road map to become debt free and partnership agreements with the neighbors. Their cows eat our grass, and we get to share in the beef. Our bees produce honey from our gum trees, and we share the treat with them.

The trains will start to get busier and life will resume, but part of me -- and part of all of us -- will long remember the disruption wrought by Covid 19. We will never be quite the same, no matter how normal life gets. And we will continue to have questions about the future. In a team meeting the other day, I high fived a team member for bringing in work. We then had to go find the hand sanitizer, momentarily forgetting the rules around contact. My data is now being monitored by an app that tracks the Bluetooth technology of every person I come into contact with in less than 1.5 meters. Some may view it as intrusive, but I trust the system will be used to help contain any new outbreaks and keep us safe. I hope it works. I do not know whether I will see my family for my 40th birthday this year, or when I will see them again in person. I do not know when or if air travel will return to normal, but I do know they are alive and well. I do not know if we will find a vaccine for the virus. We must wait and hope and create a future for all of us, whether a vaccine emerges or not.

American writer Robert A. Heinlein coined the term

"grok" in his 1961 science-fiction novel *Stranger in a Strange Land.* It means "understanding intuitively or by empathy, to establish rapport with" and sums up the feeling so many of us have now that we're sharing in this truly unique experience together. Thank you to one of my graduates for teaching me this new word. The whole world is channeling through the same emotions: fear, grief, gratitude, hope. This will change how we do business in the future. We now have a social responsibility, built from this shared empathy, to help everyone we can. It's no longer acceptable to work in a siloed way; we must work together to make a difference for the company, for our communities, for our world. We have a responsibility to support local, to buy a coffee at the station, to give to others and to keep local economies going. Social responsibility is not only about doing the right thing; it's about being a person with integrity, doing the right thing even when others might not like it. A leader who shows heart and emotion, who's motivated by grok, is a strong leader -- and many leaders have shined in these times. Unfortunately, too many others have shown a lack of grok that's caused more pain and suffering during a period of already unimaginable grief. I wonder what the future holds for those leaders, the ones who acted selfishly and abhorrently in this crisis. I wonder if their workers and citizens will hold them accountable?

There are many things we just don't know the answer to yet. But I do know one thing: We will triumph over this virus; we will come out the other side. We will feel grateful every day, look to the stars and thank them for keeping us safe and well. And we will be mindful of those who were not so lucky, for the families who lost loved ones, for the people struggling to find their normal after The Pandemic. We will look after them with empathy, with grok and with a renewed understanding of how we all really are in this thing called life together.

TAKE ACTION

You don't have to go back to the way you were before the pandemic. What things in life do you want to keep and what things would you like to change after this experience? Spend some time reflecting and make a list now.

Chapter 22

Time to Change the World

"You've always had the power my dear, you just had to learn it for yourself."

-- Glinda the Good Witch, "The Wizard of Oz"

One of the most beautiful movies we watch as children is *The Wizard of Oz*, a story that reminds us of the potential we all have waiting inside. Dorothy and her companions are looking for a leader to fulfill their desires: a way home, a brain, a heart, courage. Yet instead of finding what they seek in the Wizard, they're told to look inside. What they desire has been there all along. These characters remind me of the leadership model I've described in this book. The Lions, with their courage, must call on the intelligence of the Scarecrows -- both human and robotic-- and tap into the empathy of the Tin Men to find a way forward. While intelligence and heart are important traits in a leader, courage is key. There are times all of us feel like the Cowardly Lion, when courage ebbs and fear flows. In those moments when we feel frozen by fear, unsure where to turn next, we need to remember the story of the *Wizard of Oz* and tap into the courage inside us. Sometimes courage is as simple as speaking up, saying something when you see something happening that isn't quite right. And sometimes courage requires great sacrifice, of a job, a career, even your life. As the Lion learns, having courage doesn't mean being fearless; it means doing the thing anyway because I assure you all leaders get scared. This can be exhausting and overwhelming sometimes, especially when we're in a land of unknowns and uncertainty. But every time we cross that boundary of fear and act anyway, it gets a little bit easier to do it the next time.

Throughout this book I have called on you to confront your fears about your own shortcomings, about the unknowns of the future, about changing times and technologies. I've asked you to look inside and find the way you can lead, authentically and courageously. I've talked about my own path through the depths of spatial to the field of technology, the challenges I faced in this male-dominated space, the successes I've had and the losses I've suffered. I've told you about my visions for the future, about the technologies that have the potential to truly change the way we live and work, the way we care for each other and the way we care for our environment. And, finally, I've told you how *I* think successful leaders of the future will lead -- with heart, with passion and a little bit of a rebel spirit.

History has shown us that the greatest crimes are not committed by the people breaking the rules but by the people following the rules. I myself have been guilty of this in situations where, I later realised, I was being too honest, too too trusting, too cautious, too humble, too well … something. What happens when we decide to do what our instincts demand instead? What happens when we're comfortable with our feelings and decisions, when we lead with heart and treat others as individuals. We know technology can change the world, we've seen that happen. But it can also create disruption and sorrow if change is not implemented to the best of our ability and for the good of as many people as possible. "It's kind of bittersweet," said Ukranian-American comedian Yakov Smirnoff. "The human spirit is not measured by the size of the act, but by the size of the heart."

Fear is a powerful deterrent, and many bad things happen when scared people act or watch in silence. I'm learning that even if I do not have the courage to speak up in the moment, I must not be afraid to speak up later. This book is my own indulgent way of speaking up, challenging my fears and, hopefully, empowering leaders of tomorrow to do the same. I've been inspired by those people who shy away from the limelight at first but still make a tremendous impact. For example, Satoshi Nakamoto, the pseudonymous person who developed Bitcoin. Or the female code-breakers who worked at England's Bletchley Park doing top-secret and vitally important work during World War II. Those female heroes had to keep quiet about their successes for more than 50 years. There are numerous people throughout history who quietly effect change, from the countless unnamed men and women who fight for social justice by protesting wrongdoing to the now-famous female NASA mathematicians who got John Glenn to space, Katherine Johnson, Dorothy Vaughan and Mary Jackson (Perhaps my children will grow up watching the movie about these women, *Hidden Figures*, the way I grew up with *The Wizard of Oz*). There are courageous people in history who work hard behind the scenes on innovative products and new approaches. There are people who influence their universities, colleges, schools and communities with compassionate, quiet leadership. You don't have to be loud to be a leader -- but you do have to know when to speak

187

up. I know speaking up can be very hard to do, especially when you are in the minority or you are different from your peers. Speaking with thought, reserve and with heart is even harder.

Recently I had to speak up when I encountered an unconscious bias in an Artificial Intelligence (AI) being developed by a university. I was playing with and testing the product and encountered a gender-biased response to one of my questions. The AI obviously didn't expect to interact with a female in the industry. I took a screenshot of the misogynistic answer and let the developers know. They were shocked but took action and changed the response. I was pleased by the correction but a bigger question emerged, "Why did that response appear in the first place? Why was the AI developed with misogynistic tendencies?" As we in the technology space develop the tools of the future -- AI and other intelligences -- we need to be aware of our own biases and take extra caution to make sure they don't find a way into the products we create. We all have them, myself included (I talk about my own digital bias earlier in the book). If we as users encounter situations like the one I described, we need to ensure we speak out en masse so the issue can be corrected and prevented from happening again.

Showing courage everyday can be exhausting, but having

people around you who support you, trust you and love you no matter what can keep you recharged, energised and ready for more. Find these people. They may be part of your family, your school or communities, church or sports club, coworkers or mentors in the workplace. My circle has been crucial to my success, and you will find the same. Look for people you can trust, you can confide in. Tell them what you want to do, what your greatest ambitions are and what you think you can achieve. Ask them to hold you accountable, to remind you of your dreams when your courage wanes. Ask them to be part of your journey. When I was a kid, my teachers and parents were used to me having big ideas that would sometimes spill out from my little brain fast and furiously. They would tell me to slow down and break it down for them so they could understand what I was talking about and how they could help. That process still helps me today. I break down big projects, big goals, even this book into small parts that I can work toward on a daily basis, and you can do this too. Anyone can change the world one small act at a time. Just remember to reward yourself when you achieve part of your goal. Celebrate when you are courageous, when you are a lion or lioness. Celebrate the little successes of life.

For the female leaders reading this book, I hope the

journey is a bit easier for you than what I experienced and the women that went before me. I've worked in this male-dominated industry for almost 20 years, and I must say that being a female technologist has its benefits and drawbacks. On the whole though I have absolutely loved 99% of the time, and my male and female mentors have been invaluable to my journey. I've had wonderful men act as sponsors, coaches, teachers and mentors throughout my career. My work in the industries of engineering, resources, mining, utilities, the environment and the government has been challenging and rewarding. And all of these experiences, including my expertise in data and spatial fields, have contributed to this story, *my* story. They've taught me to be resilient and made me grateful for those who paved the way. I hope the bricks I've laid down along this path keep you steady and help you walk even stronger, even further -- much farther than you could imagine. My last bits of advice? Stop to enjoy your success and don't get too carried away with life. Have fun, show heart, wear your mistakes and missteps like a badge of honour. Find the magic that comes from doing what you love. Above all, dream and dream big! You can never dream big enough. When you wake to a dream achieved, sing another song and dance to the vision of what you'd like to achieve next. Dream again and dream bigger than you dreamt before. I believe we are

1,000 times more capable than we think. We are 1,000 times more

courageous than we sometimes feel. I truly believe we have the

ability to change the world -- and I'm believing in you to do it.

MUSIC TO FEED OUR SOUL

"If music be the food of love, play on, give me excess of it that, surfeiting. The appetite may sicken and so die"

William Shakespeare 12th Night Act 1, scene 1, 1-3.

For those who have known me for years, will know that for me Music is part of life. I will often work with earphones in, in order to focus and get work completed. Music for me is a part of being in the zone. In my younger years I would attend music school on a Saturday with my sister. Fond memories.

Studying music both theory and practical gave us an understanding of Maths and patterns. It also gave me an outlet and the opportunity to be creative when composing Music.

Whilst writing this book, I found that I was my most creative when listening to music. I have listened to music almost every time I have been writing. I thought it might be nice to share with you some artists. I would love to hear what music you listened

to whilst reading this book, drop me a tweet, post or email and let

me know.

Adele, Queen, Delta, Stereophonics, Jack Johnson, Coldplay, Muse,
Dean Lewis, Green Day, Snow Patrol, Robbie Williams, James
Bond, Kaiser Chiefs, Marvel Music, Dan Sultan, Disney theme
tunes, 80's Songs, Christmas Music Mix. Many movie Soundtracks.

ABOUT THE AUTHOR

Sarah James has over 20 years' of experience in data and its use within technology and engineering. She has a passion for understanding and solving clients 'problems leveraging Analytics and Data to do so. Sarah started her career in Geographical Information Systems and moved to Technology and Data delivering innovative projects that had never been done before. Sarah's experience enables her to help our clients maturing and transforming with the future in mind by developing innovative strategies, roadmaps, clearly articulating business value and leading teams to deliver the best solution.

Sarah is passionate about Women In Technology and Diversity of Thought (Neurodiversity).

Sarah's mission is to inspire others and to help her clients with new ideas and initiatives, leading them to change/transform in a way that creates innovation and ensures the data is always considered.

Sarah is Welsh born and grew up in Oxfordshire, UK. Sarah studied at Plymouth University Geography and at University of New South Wales Masters of Science and Technology in Geographic Information Systems.

Sarah with her Husband Robert own an off-grid sustainable farm in Western Australia. Sarah is mum to two beautiful girls.